Dielectrics

Dielectrics

P. J. HARROP, B.Sc., Ph.D., F.Inst.P.,
Product Development Manager,
Plessey Capacitors,
Bathgate, West Lothian

LONDON BUTTERWORTHS

THE BUTTERWORTH GROUP

ENGLAND
Butterworth & Co (Publishers) Ltd
London: 88 Kingsway, WC2B 6AB

AUSTRALIA
Butterworth & Co (Australia) Ltd
Sydney: 586 Pacific Highway, Chatswood, NSW 2067
Melbourne: 343 Little Collins Street, 3000
Brisbane: 240 Queen Street, 4000

CANADA
Butterworth & Co (Canada) Ltd
Toronto: 14 Curity Avenue, 374

NEW ZEALAND
Butterworth & Co (New Zealand) Ltd
Wellington: 26–28 Waring Taylor Street, 1

SOUTH AFRICA
Butterworth & Co (South Africa) (Pty) Ltd
Durban: 152–154 Gale Street

First published 1972

ISBN 0 408 70387 3 standard
0 408 70388 1 limp

Filmset and printed in England by
Page Bros (Norwich) Ltd, Norwich

Preface

This book is about insulators, otherwise known as dielectrics, a poorly defined yet major class of materials that conduct virtually no electricity under low d.c. electrical fields. The approach has been one that employs the minimum of mathematics and electromagnetic theory to establish the underlying principles involved, yet includes the major advances of the last five years. The author has tried to visualise a reader who is an HNC student or first or second year undergraduate in Materials Science, Applied Physics or Electrical Engineering. He also hopes that the book will be of some use to students of other disciplines who wish to grasp some of the more recent ideas and practice in this subject. The general reader should be able to avoid the mathematical sections and still get a coherent picture of the whole. An introduction to the more advanced aspects of the subject is provided by reading lists at the end of each chapter and appendixes.

The insidious effects of water, sodium contamination and the like that must concern anyone who ventures beyond the vector diagram and into the real world have been detailed. Conversely, the author has tried to resist the temptation of the physicist to write at length on ferroelectrics, recognising that they represent only a small proportion of the dielectrics in use today.

SI units are used except for miscroscopic energies where the more commonly employed unit eV is quoted. This is internationally recognised for use with SI units. There is a certain confusion of terms in the literature and it must be admitted at the outset that the definitions laid down of such terms as 'paraelectric' and 'polar material' are merely those that are considered to be the most widely accepted.

The author is most grateful to the many people who have commented on the text, in particular Dr. B. Salvage and Dr. A. R. Morley, and also to his wife for shorthand and typing.

Contents

Main symbols used

α	polarisability, Townsend's first coefficient
A	area, constant
β	distribution function
B	constant
C	capacitance
C_0	constant
C^*	complex capacitance
C_S	equivalent series capacitance
D	flux density, diffusion coefficient
d	thickness
δ	loss angle
ε'	relative permittivity
ε''	imaginary part of complex permittivity
ε^*	complex permittivity
ε_0	permittivity of free space
ε_s	permittivity at relatively low frequency
ε_∞	permittivity at relatively high frequency
e	charge on electron
E	electrical field
E_B	breakdown field
E_i	internal field
ϕ	work function
F	force, temperature coefficient of volume polarisability
G	conductance
γ	Townsend's second coefficient

γ_c	temperature coefficient of capacitance
H	constant
I	current
i	current variable
j	$\sqrt{-1}$
J	current per unit area
k	Boltzmann's constant
K	constant
λ	linear expansion coefficient
l	distance
μ	mobility
M	molecular weight
m	dipole moment, particle mass
N	number
n	number, refractive index
N_A	Avogadro's number
η	viscosity
P	polarisation, pressure
q	elemental charge
Q	energy
r	distance
ρ	density
R	resistance, gas constant
$R_{d.c.}$	d.c. resistance
R_E	electrode resistance
R_S	equivalent series resistance
σ	conductivity
S	constant
t	time
T	temperature
τ	time constant
θ	phase angle, characteristic temperature
V	voltage, volume, breakdown voltage
$\overline{V^2}$	mean square velocity
ω	angular frequency
x	distance
z	number of increments of charge
Z	impedance

1

Background

In the 1950s and 1960s, the student of the electrical properties of materials tended to receive a large treatise on metals, followed by a large treatise on semiconductors and only the briefest sketch on dielectrics. There were historical reasons for this. Semiconductors was a fast-moving and theoretically sophisticated field; exploitation of new phenomena and materials could be seen on all sides.

By contrast, the field of dielectrics had moved very little in twenty years apart from the early examination of piezoelectric and ferroelectric phenomena. Although they were extremely widely employed, this was on an empirical basis. Capacitors, cables, motors and transformers used natural products like mica, paper and common oils, conceptually only a moderate advance on the tar, grease, rope and gutta percha used in the last century. Recipes for ceramics and varnishes had been handed down between generations. Even the simplest information, such as the mobility of the few charge carriers remaining in any real insulator, was usually undetermined. Yet it was these few charge carriers that usually limited their use in engineering.

Paradoxically, all this was largely changed by the new electronics of the 1960s—microcircuitry. Increasingly it became necessary to stress general insulation to a point dangerously close to breakdown, and the heart of the 'active' device often became an insulating 'gate' rather than a simple semiconducting junction. Tightened specifications for electronic and electromechanical components could no longer be met using impure minerals or paper compositions. Dielectrics, as much as semiconductors, had to be 'tailor made' for the improved devices demanded by the customer. This called for a closer look at relevant materials which itself led to the identification of new dielectric phenomena that could be exploited. Piezoelectrics are now widely used: electrets and electro-optic materials soon will be. These are a far cry from plain insulation.

In parallel with this, the 1960s saw the physicist making great strides in the understanding of the gaseous state. Since the new electronics often called for glassy materials, there was considerable interest in conduction in disordered solids, leading naturally to the study of conduction in liquids—previously the sole province of the chemist—and to the invention of switches and diodes based on non-crystalline solids.

Of course, electronics itself was by now accelerating at a remarkable pace. Computers and colour television were the tip of the iceberg. The housewife's use of electric motors and switches for every purpose in the home called for complex control circuits and interference suppressors: the motor car became a haven of electronic devices; trains and aircraft used ever more electronics. In parallel with this, the increase in demand for electricity led to unprecedented development of power networks. Most of the components needed had to be made cheaper and more reliable, which meant smaller. The dielectrics involved had to be worked much nearer their limits.

The result has been that insulating materials, or 'dielectrics' (we use the terms interchangeably) are now a sophisticated field of study in their own right. The construction of a transformer or a transducer can consist of synthetic materials designed from basic principles. One can now learn the basic theory of metals on the one hand and that of dielectrics on the other. This covers most known materials. Semiconductors, those comparatively rare and peculiar materials in between, can then be understood in context.

1.1 COULOMB'S LAW

We now recall some simple electrical theory and present the mathematics that will be useful when looking at practical dielectrics.

Coulomb's law is expressed by the equation

$$F = \frac{q_1 q_2}{4\pi\varepsilon'\varepsilon_0 d^2} \tag{1.1}$$

where F newtons is the force between two charges q_1 and q_2 coulombs separated by a distance d metres, ε' being a constant of proportionality defined as the relative permittivity (dielectric constant) of the intervening matter. For a vacuum, ε' is chosen to be unity. ε_0, the other constant, is called 'the permittivity of free space'. ε_0 balances the equation dimensionally and has a value of $(1/36\pi) \times 10^{-9}\ \mathrm{Fm}^{-1}$.

The relative permittivity is commonly abbreviated to permittivity. It is never less than unity. For gases ε' is usually within a few per cent of 1. Most solids have values below 40 but a few run into tens of thousands. Liquids commonly have values of less than a hundred.

1.2 A PERFECT DIELECTRIC

It can be shown that the capacitance C farads, of a parallel plate capacitor, neglecting edge effects, is given by

$$C = \frac{\varepsilon'\varepsilon_0 A}{d} \tag{1.2}$$

where A m^2 is the area of electrodes, d m the thickness. This device will obey Ohm's law under d.c. fields but under a.c. it will impede the flow of current by an amount that varies with frequency f Hz (one Hz is one cycle per second). One therefore defines the a.c. equivalent of resistance $R\ \Omega$ as the 'impedance' $Z\ \Omega$. In the simplest case of a perfect dielectric $R = \infty$ and

$$Z = \frac{1}{\omega C} \tag{1.3}$$

where $\omega = 2\pi f$.

1.3 A REAL DIELECTRIC

Any real dielectric conducts electricity to some small extent when a d.c. voltage is placed across it. To describe this situation we define our terms by considering a *parallel* combination of resistance R and capacitance C as in Figure 1.1 where the capacitance C has the same nature as the perfect capacitor considered above and the conductance G represents an ideal resistor, i.e. one having no

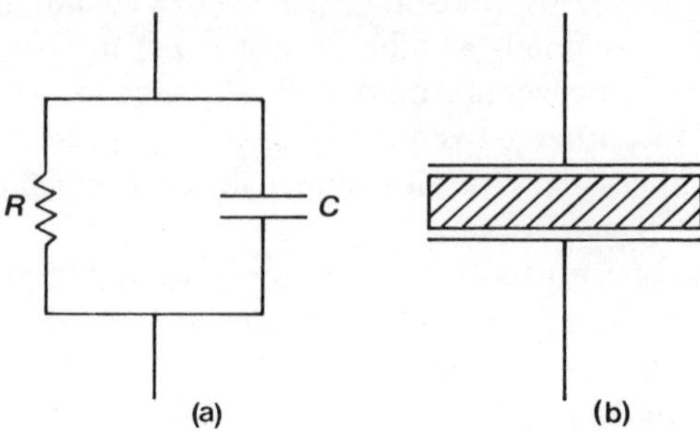

Figure 1.1. The defined equivalent circuit (a) of a dielectric (b) sandwiched between two metal plates for measurement purposes, this arrangement being known as a capacitor

capacitive or inductive parts. Clearly then, the d.c. conductance G ohms^{-1}, of our perfect dielectric is given by

$$G = \frac{\sigma A}{d} = \frac{1}{R} \tag{1.4}$$

for the parallel plate configuration, neglecting edge effects, where σ is conductivity ($\text{ohm}^{-1}\ \text{metre}^{-1}$), A m^2 the area of electrodes, d m electrode separation and R Ω the resistance. As above, the capacitance C F is related to the geometry by

$$C = \frac{\varepsilon' \varepsilon_0 A}{d} \tag{1.5}$$

This is very simple when we consider a real dielectric subjected to a d.c. potential, but when an a.c. voltage waveform is applied to such a material the current induced does not reach maxima and minima at the same time as the applied voltage does—it 'leads in phase' the potential applied. It is convenient to use a vector diagram to describe this, with real and imaginary parts as in Figure 1.2. The angle θ (the 'phase angle') then represents the angle by which the current waveform leads the voltage waveform. Thus if G is zero as with an ideal capacitor, $\theta = 90°$; and if $C = 0$ as for a perfect resistor, then $\theta = 0$. Using complex notation, where $j = \sqrt{-1}$, the impedance Z of the real dielectric is given by

$$\frac{1}{Z} = \frac{1}{R} + j\omega C \tag{1.6}$$

To describe this it is convenient to define a generalised 'permittivity' that has a real and imaginary part and will include resistive as well

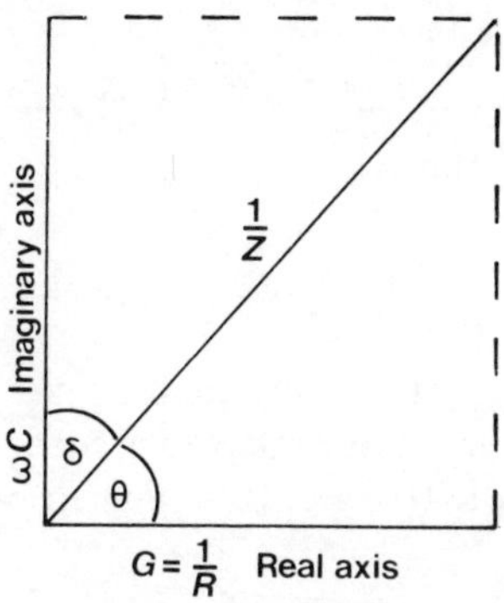

Figure 1.2. Vector diagram of the electrical response of a dielectric

as capacitive contributions. Note that this is not a question of proof. We are simply defining terms in a way that will be convenient later. Now write

$$\varepsilon^* = \varepsilon' - j\varepsilon'' \tag{1.7}$$

where ε' is the relative permittivity as used before and $j\varepsilon''$ is associated with the resistive vector. Using a little algebra, expressions are obtained for ε'' and a very useful quantity, the 'loss tangent' $\tan \delta$.

By analogy with an ideal capacitor, one can express the 'complex capacitance' C^* of a real slab of dielectric as

$$C^* = \varepsilon^* C_0 \tag{1.8}$$

where $C_0 = \varepsilon_0 A/d$ for a parallel plate capacitor, i.e. it is merely a geometrical constant, and the impedance Z is given by

$$Z = \frac{1}{j\omega C^*} = \frac{1}{j\omega C_0(\varepsilon' - j\varepsilon'')} \tag{1.9}$$

describing both the resistive and the capacitive parts in a consistent fashion. Now, from equation (1.6)

$$Z = \frac{R}{1 + j\omega CR} \tag{1.10}$$

Hence, combining equations (1.9) and (1.10) and equating real and imaginary parts, $\varepsilon' = C/C_0$, the relative permittivity as expected, and

$$\varepsilon'' = \frac{1}{\omega C_0 R} = \frac{\sigma}{\omega \varepsilon_0} \tag{1.11}$$

It is important to note that

$$\frac{1}{\omega RC} = \frac{\varepsilon''}{\varepsilon'} \tag{1.12}$$

$\varepsilon''/\varepsilon'$ is defined as being equal to a quantity tan δ. (Note this is for *any* material of *any* geometry.) By reference to Figure 1.2, this definition is consistent with $\delta = 90° - \theta$ as drawn. Thus

$$\tan \delta = \frac{\varepsilon''}{\varepsilon'} = \frac{1}{\omega RC} = \frac{\sigma}{\omega \varepsilon' \varepsilon_0} \tag{1.13}$$

and again ε' is the conventional relative permittivity.

Let us now consider some general features of a real dielectric. Conduction through a resistor, unlike a perfect capacitor, must always cause joule heating and the equivalent circuit defined above is no exception. From consideration of Figure 1.2, sin δ is a measure of the resistive impedance and hence the heat dissipated or electrical power absorbed. Since this book is concerned with materials with very little conduction, e.g. tan $\delta < 0{\cdot}5$, one can usually write sin $\delta \approx$ tan δ and therefore the energy absorbed is proportional to tan δ.

Accordingly, if one has a device for measuring the tan δ of a material as a function of the frequency of the field applied, one is carrying out a form of spectroscopy. It is simply measuring the energy absorbed as a function of frequency of electromagnetic radiation for frequencies typically between 1 Hz (1 Hz = 1 cycle per second) and 10^9 Hz. Spectroscopy in the visible range refers to measurement of the absorption of electromagnetic radiation as a function of frequency at frequencies around 10^{15} Hz. Such spectra are often quoted in the sales literature for new dielectrics as they are useful in design.

Sometimes $\sin \delta$ is referred to as the power factor and $\tan \delta$ as the dissipation factor, dielectric loss or loss tangent of the material in question. Later it will be shown that it is occasionally easier to determine ε'' than $\tan \delta$.

It is most important to realise that although the conductivity of dielectrics is twenty to thirty magnitudes lower than that of metals it still limits most applications. This book therefore takes careful note of the residual conductivity mechanisms.

We now consider some practical quantities that will be needed later. ε' is a quantity that varies only slowly with variables such as temperature and pressure and it is often intrinsic to the material in question. Typical values were indicated in the earlier discussion. In contrast, σ, and hence ε'', varies sharply with temperature, pressure and other variables and is usually extrinsic, being a function of trace impurities, structural weaknesses, and so on.

It is useful to define an electrical flux density D by analogy with magnetism, where

$$D = \varepsilon' \varepsilon_0 E \tag{1.14}$$

D is therefore a vector in the direction of field E, being a measure of the extent to which the surroundings modify the field. D is expressed in C m^{-2} and is sometimes known as the electrical displacement. It is numerically equal to surface charge per unit area.

1.4 RESISTIVE PROPERTIES OF REAL DIELECTRICS

1.4.1 MOBILITY

To analyse the low, but limiting, conductivity of a practical dielectric one defines a measure of the agility with which any conducting species moves under a potential, by writing

$$\sigma = ne\mu \tag{1.15}$$

for singly charged species where σ is the conductivity due to n charge carriers and e is the value of a single charge (the charge on an electron), being $1{\cdot}62 \times 10^{-19}$ coulombs. The mobility μ is therefore a constant of proportionality defined by equation (1.15). Typical values of mobility for dielectrics lie between 10^{-4} and 10^{-14} $m^2\ V^{-1}\ s^{-1}$ at 20°C, and increase with increase in temperature, whereas μ for metals and semiconductors usually decreases with increase in temperature. Values for metals and semiconductors are usually between 10 and 10^{-4} $m^2\ V^{-1}\ s^{-1}$. There will be no need to consider σ and μ as vectors in this book.

1.4.2 FREQUENCY OR TIME VARIATION OF CONDUCTIVITY

In order to appreciate the basis of the analyses of the frequency and time response of real dielectrics given later in the book, including the so-called Debye equations, it is useful to understand the characteristics of common rate processes.

Most physical and chemical processes obey the simple rule that the rate of change of the quantity n considered is proportional to n, i.e.

$$\frac{\mathrm{d}n}{\mathrm{d}t} = Kn \tag{1.16}$$

where K is a constant. This implies that

$$\int \frac{1}{n}\,\mathrm{d}n = K \int \mathrm{d}t \tag{1.17}$$

$$n = A \exp Bt \tag{1.18}$$

where A and B are constants. It will therefore come as no surprise that most of the kinetic equations in this book include exponentials or the converse, logarithms.

A good example of a time exponential is the concept of *time constant* used in this book. A dielectric has a pure d.c. resistance R and capacitance C defined by the equivalent parallel components. If the ideal capacitance C in this equivalent circuit supports a charge Q and a voltage V_c at time t, it will discharge through the resistive part R. However, $V_c + V_R = 0$, where V_R is the voltage across the resistance R. Now $V_c = Q/C$ and $V_R = (\mathrm{d}Q/\mathrm{d}t)R$ by definition, so that

$$\frac{Q}{C} + \frac{\mathrm{d}Q}{\mathrm{d}t} R = 0$$

Integrating

$$\log_e Q = \frac{-t}{RC} + K \tag{1.19}$$

where K is a constant. If $Q = Q_0$ at $t = 0$, $K = \log_e Q_0$ and

$$Q = Q_0 \exp(-t/RC) \tag{1.20}$$

The quantity RC is defined as the time constant, τ, of the dielectric, being the time for a charge to decay 1/exp of its original value with the dielectric on open circuit. In practice τ can be taken as the product of the low field d.c. resistance of the dielectric and the capacitance at, say 1 kHz. Since C is a relatively invariant quantity for most dielectrics, variation of τ with temperature, pressure, etc., is usually dominated by variations in R. Values of τ for most dielectrics lie between seconds and months.

1.4.3 TEMPERATURE VARIATION OF CONDUCTIVITY

There now follows the background of why the conductivity of dielectrics varies exponentially with temperature. If one considers an ideal gas of low density with the particles of mass m undergoing fully elastic collisions with each other, it emerges that their mean energy Q is proportional to temperature. In fact

$$Q = \frac{1}{2} m\overline{V^2} = \frac{3}{2}\frac{R}{N_A} T \tag{1.21}$$

where $\overline{V^2}$ is the mean square velocity, R the gas constant and N_A Avogadro's number.

Actually, it is convenient to put

$$\frac{R}{N_A} = k \tag{1.22}$$

where k is Boltzmann's constant (see Appendix 4 for values of constants), i.e.

$$Q = \frac{3}{2} kT \tag{1.23}$$

This result, although derived for an ideal gas, is applicable to many diffusing species in liquids and solids also. At 20°C, kT has a value of 0·023 eV, where 1 eV is equal to the amount of energy gained by an electron passing through 1 V potential difference ($1 \text{ eV} = 1{\cdot}92 \times 10^{-19}$ J).

Now if the number of particles having energy between Q and $Q + \mathrm{d}Q$ is $N(Q)$ ($\mathrm{d}Q$), the probability of a given particle having this energy can be derived as

$$F(Q)\,\mathrm{d}Q = \frac{N(Q)\,\mathrm{d}Q}{N} = 2\pi \left(\frac{1}{\pi kT}\right)^{\frac{3}{2}} Q^{\frac{1}{2}} \exp\left(-\frac{Q}{kT}\right) \mathrm{d}Q \quad (1.24)$$

This is called the Maxwell–Boltzmann function. Note that the exponential dominates variability of $F(Q)\,\mathrm{d}Q$ with temperature. The form of the function is shown in Figure 1.3.

By consideration of the Maxwell–Boltzmann function it is obvious that when a perfect gas particle is faced with an energy

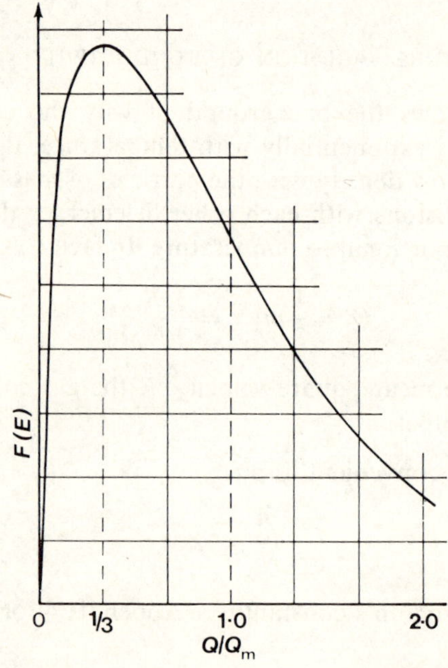

Figure 1.3. Maxwell–Boltzmann distribution function where Q_m is the mean energy

barrier Q' which prevents it from moving to a new position, the probability of its surmounting the barrier, and hence the number of particles n passing over the barrier per unit time, is approximately given by

$$n = S \exp\left(-\frac{Q'}{kT}\right) \tag{1.25}$$

where S is a constant. This is known as the Arrhenius equation. It leads one to plot the logarithm of n against $1/T$. This gives a linear relationship, the slope of the line giving Q'.

For dielectrics the parameter n in equation (1.25) may for instance be the number or mobility of conducting ions or electrons. Both can vary exponentially with temperature and in the conductivity equation one will dominate. More detailed consideration makes one allow for a dependence of S on T, the simplest cases giving $S \propto 1/T$. One can therefore plot log σT against $1/T$ for ions in crystalline solids for example, and hence get a more meaningful value of Q' from the slope.

1.4.4 BREAKDOWN

Of course, the highest voltage a dielectric will withstand is of interest in the design of equipment. As the potential across any material is increased, Ohm's law is obeyed followed by a region of non-ohmic behaviour. Finally a potential is reached at which the current runs away in an uncontrollable manner. This field is known as the breakdown field of the substance. It is also quoted as the breakdown strength, electric strength or dielectric strength. Insulators typically break down at 10^6 to 10^9 V m^{-1} d.c. at 20°C. This varies considerably with the method of measurement. A high vacuum has the highest electric strength followed by thin homogeneous solids and then liquids ($\leqslant 2 \times 10^8$ V m^{-1} with typically best purity).

1.5 CAPACITIVE PROPERTIES OF REAL DIELECTRICS

1.5.1 DIPOLE MOMENT

In understanding the magnitude of C for a given material we must identify the various charged species contributing to the total charge on the capacitor. When the capacitor is charged, the dielectric in it is said to be polarised. Usually, the polarising species approximates to a dipole, that is electrical charges of equal and opposite sign, e.g. $\pm e$ separated by a distance d. This dumb-bell-like configuration can rotate, and to calculate the effects of this, we define the dipole moment as

$$m = ed \tag{1.26}$$

where m is a vector quantity, taken along the dipole, as sketched

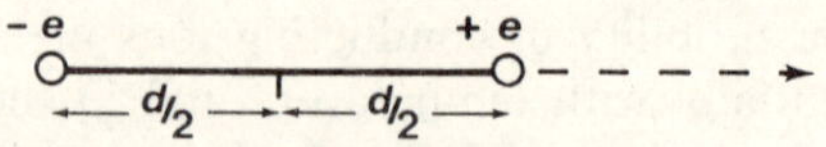

Figure 1.4. A simple dipole, showing the defined direction of electric moment

in Figure 1.4. Various types of dipole found in dielectrics will be described in Chapter 2.

1.5.2 POLARISABILITY AND POLARISATION

An electrical field E_i orienting, or creating dipoles is referred to as polarising and the polarisability α_d of a given dipole is defined by

$$m = \alpha_d \boldsymbol{E}_i \tag{1.27}$$

where m and $\boldsymbol{E}_i$ are vectors. Note that $\boldsymbol{E}_i$ is the field acting *at* the dipole. $\boldsymbol{E}_i$ does not necessarily equal the field at the outside of the material in question. We shall discuss this further later.

Macroscopically, the polarisation P is taken as the difference between the bound surface charge measured and that for vacuum, i.e. using the concept of electric flux density D given earlier,

$$P = D - \varepsilon_0 E \tag{1.28}$$

This is therefore a vector quantity having the direction of D whence,

$$P = (\varepsilon' - 1)\varepsilon_0 E \tag{1.29}$$

P is equal to the electric dipole moment per unit volume induced in the substance by the applied field.

1.5.3 TEMPERATURE COEFFICIENT OF CAPACITANCE

Dielectrics are used at all temperatures and the temperature variation of capacitance may be crucial in a given device. An important parameter used to indicate the variation of capacitance with temperature is the temperature coefficient of capacitance, γ_c, which is defined by

$$\gamma_c = \frac{1}{C}\left(\frac{\partial C}{\partial T}\right)_P \tag{1.30}$$

where T is the temperature and P the pressure (constant). γ_c is usually expressed in parts per million (ppm) per degree Celsius since the best dielectrics have values between -400 and $+400$ ppm/°C.

1.5.4 ENERGY STORED

Dielectrics store considerable amounts of energy when a field is applied to them, and capacitors are used to power spot welding machines and photographic flash equipment in exactly the same way as their mechanical equivalent, a spring, powers a pop gun.

The energy stored by a dielectric of capacitance C after an infinite time at voltage V can be found as follows

$$\text{final charge } Q = CV \text{ coulombs} \tag{1.31}$$

Intermediate charge, q, corresponds to a potential difference $V = q/C$.

Suppose a small charge Δq coulombs moves around the circuit, when a charge q is present. The work done on this elemental charge in its movement against the potential V equals $V\Delta q$, i.e.

$$\frac{1}{C}\, q\Delta q \text{ joules} \tag{1.32}$$

This work represents energy stored in the capacitor. Thus the energy stored for the full charging potential V is given by

$$Q = \int_0^Q \frac{Q}{C}\, q\, \mathrm{d}q \tag{1.33}$$

$$= \frac{1}{2}\frac{Q^2}{C} \tag{1.34}$$

$$= \frac{1}{2}CV^2 \text{ joules [using equation (1.31) in equation (1.33)]} \tag{1.35}$$

Capacitors storing energy for sudden release in the electro-discharge forming of metals may provide megajoules of energy, whereas small capacitors used in electronics may only store microjoules.

FURTHER READING

ANDERSON, J. C., *Dielectrics*, Chapman and Hall, London (1964)

BLEIL, D. F., in *American Institute of Physics Handbook*, McGraw-Hill, New York (1963)

DEKKER, A. J., *Solid State Physics*, Macmillan, London (1962)

DYSON, J. E., *Introductory Physics for Electrical Engineers*, McGraw-Hill, New York (1969)

KIP, A. F., *Fundamentals of Electricity and Magnetism*, McGraw-Hill, New York (1969)

KITTEL, C., *Introduction to Solid State Physics*, Wiley, New York (1966)

ZAKY, A. A. and HAWLEY, R., *Dielectric Solids*, Routledge and Kegan Paul, London (1970)

2

The nature of matter

2.1 STATES OF MATTER

Matter is most conveniently divided into four states which generally follow increasing closeness of atomic packing, namely,

GASES, LIQUIDS, GLASSES, CRYSTALS.

Gases and liquids are well known. 'Glasses' or 'amorphous' materials are arbitrarily defined as those solids having no three-dimensional atomic ordering over distances greater than 2 nm: (atoms are tenths of an nm across). This definition includes some polymers. No solid or liquid is completely structureless.

'Crystals' exhibit long range ordering. If this ordering is consistent throughout a given solid, a rare phenomenon in nature, they are 'single crystals'. Crystalline agglomerates are called polycrystals. Examples of the above states of matter are given in *Table 2.1*.

One can also have two-phase (i.e. two-state) solids such as the glass-crystals polypropylene or porcelain, and multi-phase dielectrics are not unusual.

Note that a given substance does not necessarily pass through all phases on heating. A solid may form a gas directly ('sublimation'). Some liquids condense to a glass or a crystalline solid depending on how quickly they are cooled. However, as a general rule ionic solids—those where the ions (charged atoms) do not share electrons —do not form glasses. By contrast, covalent solids (those sharing electrons) can readily form glasses. The latter include most polymers

Table 2.1 EXAMPLES OF THE FOUR STATES OF MATTER AT 20°C. THESE EXAMPLES HAPPEN TO BE ELECTRICAL INSULATORS

Gas	Nitrogen
Liquid	Paraffin
Glass	Polystyrene, borosilicates
Crystal	Mullite (polycrystal)
	Mica (single crystal)

and silicates. Diagrammatic illustrations of typical insulating ionic and covalent solids are given in Figure 2.1.

As one might expect, the extent to which compounds are ionic or covalent is a function of their electronic orbital structure. The two

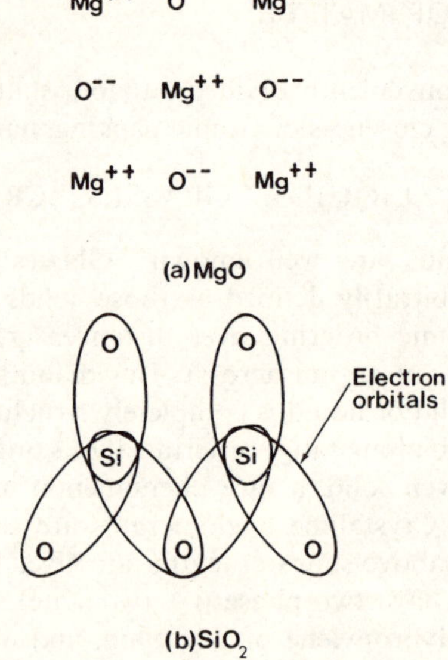

Figure 2.1. Diagrammatic illustration of (*a*) *an ionic and* (*b*) *a covalent solid*

extremes can be seen from the Periodic Table as in Figure 2.2. (the full Periodic Table is given in Figure 6.1 at the end of the book).

The group numbers are a guide to valence. Thus Group 1 atoms generally form simple non-deformable spherical univalent ions; Group 2 atoms form divalent ions, etc., until one reaches Group 4 after which the ions are particularly multivalent.

On the left are the metal ions. These are cations because, being positively charged, they are attracted to the cathode in electrochemical reactions. On the right there are Group 7 anions, which are univalent. Group 6 anions are generally divalent. Thus the simple ionic solid is the 'alkali halide'—a Group 1A cation with a Group 7B anion. Conversely, complex electron-sharing is characteristic of high valency cations and anions, most notably carbon or silicon

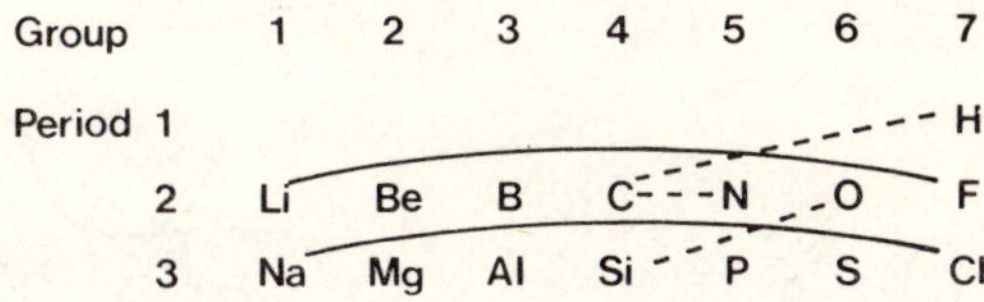

Figure 2.2. Part of the Periodic Table showing ——— ionic and ------ covalent bonds

combined with oxygen and/or hydrogen. This covers the polymers and silicates mentioned above.

Inorganic solids are composed of atomic or molecular building blocks that approximate to table-tennis balls stacked together in lines. As an inorganic substance is heated, the heating is represented by the balls jostling about. Suddenly, at a certain temperature, the balls will stop vibrating about an ordered position and will become disordered. A liquid has been formed. Similarly the attraction between the balls will be overcome as the temperature continues to rise and vibrations increase and a gas is formed. Accordingly, the transitions between the four states of matter are relatively sharp.

By contrast simple polymers may consist of chains of thousands of tightly bound atoms which repeat a basic monomer structure. The chains interact little with each other. Consequently the simplest polymers can be imagined as spaghetti. Different types of polymer structure are illustrated in Figure 2.3.

It is now understandable why polymer structures change only gradually and in a complex fashion between the states of *Table 2.1*. For instance sharp melting points are replaced by broad softening regions of temperature. The transition from solid (be it crystal or glass) to liquid is often complicated by an intermediate glass transition (the term is a misnomer) as in Figure 2.4. Electrical properties often change at the polymer's glass transition.

Nevertheless, the very independence of polymeric molecules leads to most mechanical and electrical properties being simply related

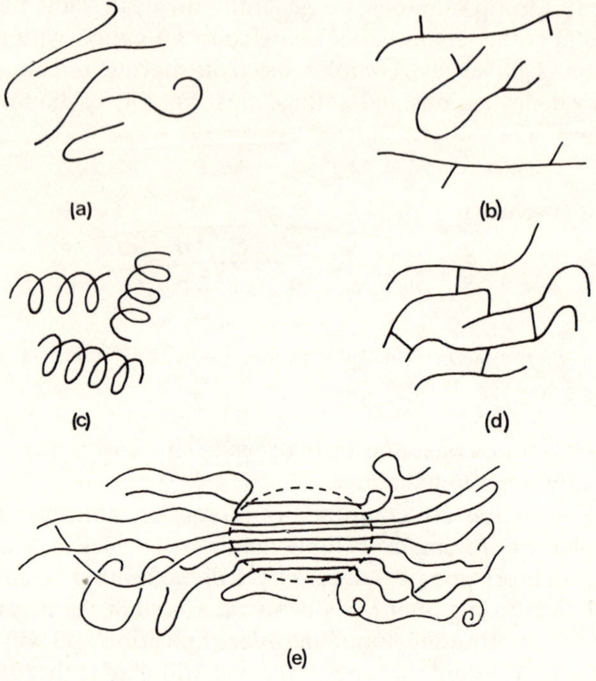

Figure 2.3. Illustration of the molecular shape of various types of polymer. (a) Linear, (b) polar (a particular case), (c) elastomer (rubber), (d) cross-linked and (e) partially crystalline linear

to structure. The same properties of inorganic compounds are more often a complex function of the whole.

The energy holding ions together in polymeric molecules, ionic solids and liquids usually varies between 5 and 30 eV (where again 1 eV is the energy gained by an electron traversing a potential of one volt, $1{\cdot}92 \times 10^{-19}$J).

2.2 TYPES OF CONDUCTING SPECIES

All dielectrics, except a perfect vacuum, pass some charge when a constant low field is applied to them. Although the currents may be very small, they more often than not limit the practical application of the materials concerned and are therefore of vital interest. The charge may only pass a small distance, in which case it is known

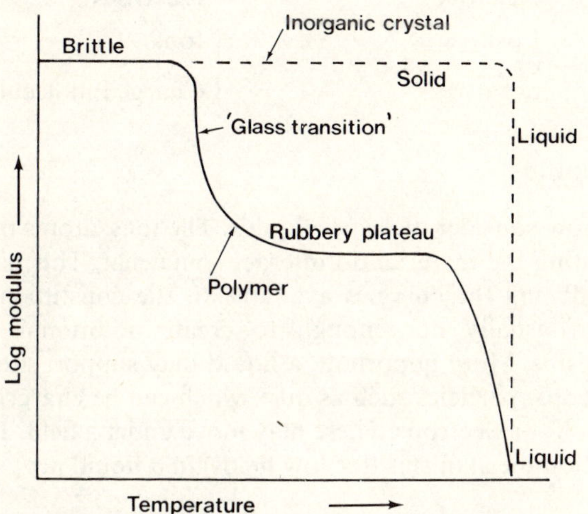

Figure 2.4. Comparison of the behaviour of an inorganic crystal and a polymer. An inorganic glass would exhibit intermediate behaviour

as polarisation, or it may pass completely through the material. This section is concerned with the latter phenomenon. We ask the question, 'What species can carry electrical charge through a dielectric?'

2.2.1 GASES

It is simplest to begin with a gaseous dielectric. Here the atoms or molecules present virtually do not interact and, since their constituent electrons are tightly bound in sharp allowed energy levels, no charge can pass through unless the system is perturbed. Such perturbation may for instance be the presence of impurity or the action of radiation such as light.

Basically the charge carried through is directly or indirectly due to electrons. The electron may move on its own or associated with an atom, constituting an ion. The charge on the ion may be positive or negative depending on whether the original atom lost or gained an electron. Further, ions or electrons may move alone or be carried on molecules (charged molecules).

The possibilities (for low fields) are:

Negative	Electrons
Positive or negative	Ions; Charged molecules

2.2.2 LIQUIDS

Let us now consider dielectric liquids. The ions, atoms or molecules constituting the material do interact somewhat. This has the effect of broadening the energies available to the constituent electrons, although usually not enough to create additional conduction mechanisms. More important, a liquid may support 'free' ions and macroscopic particles such as dust which can be charged by attraction of ions or electrons. These may move under a field. The possible conduction mechanisms (for low fields) in a liquid are:

Negative	Electrons
Positive or negative	Ions; Charged molecules; Charged particles

2.2.3 SOLIDS

Conduction in dielectric solids, both glasses and crystals, is a far more complex phenomenon than in gases or liquids. This is because, although charged macroscopic particles cannot pass (cf. liquids), there are numerous ways in which electrons, ions and charged molecules may move.

For the present we greatly simplify the situation by summarising the modes of transport in solids (under low fields) below. The meaning of the various terms will become clear after the discussion that follows.

Negative Positive	Electron Hole	{ Through allowed band { Through forbidden band
Negative or positive	{ Ion { Charged molecule	{ Interstitial { Vacancy

Electronic mechanisms of conduction in solid dielectrics

In order to understand the above electronic mechanisms, we must consider in more detail the potential energies of atomic electrons. As atoms come closer, the available electron energies broaden as in Figure 2.5. The top band is unfilled, i.e. no electrons actually

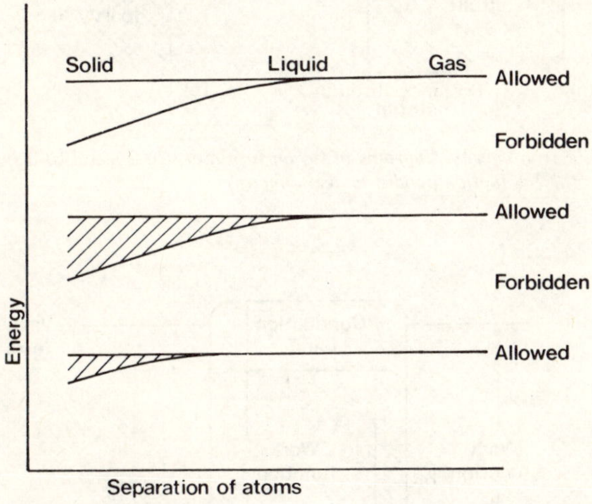

Figure 2.5. The effect of atomic separation on the energy levels available to electrons. Hatching indicates filled bands

exist within this energy range although should one be injected into it, it may move freely within it.

When one considers ways in which electrons may move in a solid, it is helpful to use an extract from Figure 2.5, known as a 'band diagram', as in Figure 2.6. Here the Fermi level E_F is a convenient energy level defined by the 'Fermi–Dirac statistics'. The probability of an electron having energy between Q and $Q + dQ$ is

$$F(Q)\,dQ = \frac{g(Q)\,dQ}{1 + \exp\{(Q - Q_F)/kT\}} \tag{2.1}$$

where $g(Q)\,dQ$ is the maximum available density of energy states in this energy range.

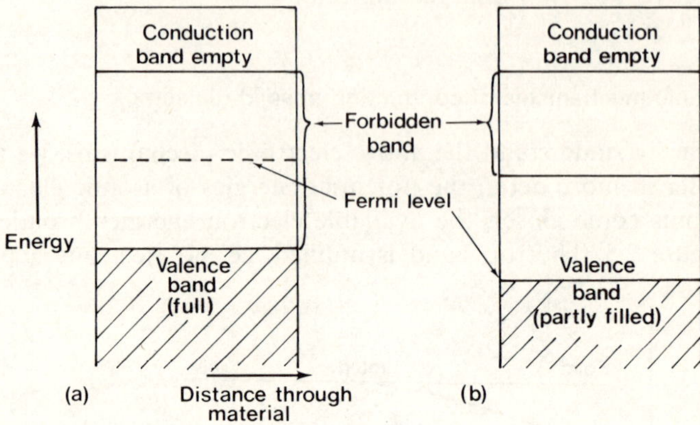

Figure 2.6. Electron energy diagrams of (a) an insulator, (b) a metal (if a full valence band overlaps the conduction band it is also a metal)

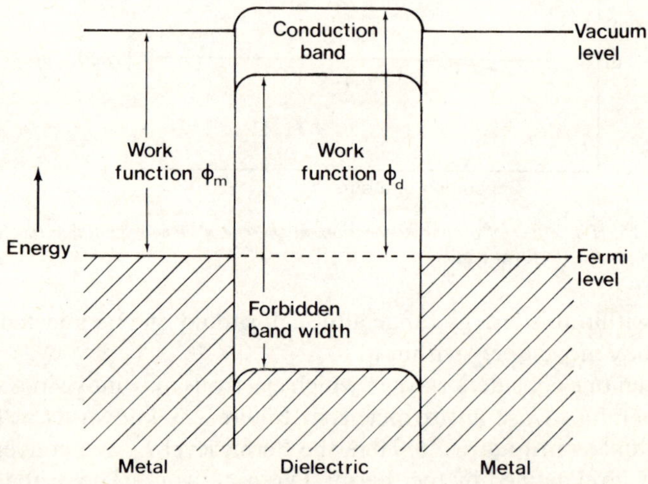

Figure 2.7. Energy diagram of an insulator with metal electrodes attached

The Fermi level of a pure perfect insulator bisects the forbidden band as shown. The Fermi level of a metal rests virtually at the highest filled energy band. This is shown for one type of metal in Figure 2.6(b). When two materials come together the Fermi levels must meet. Hence the usefulness of this concept when drawing band diagrams of complete systems.

When one measures an insulator by putting metal electrodes on it, the bands must bend as in Figure 2.7. Whether they bend up or down depends on the relative magnitudes of the work function of the metal, ϕ_m, and that of the dielectric, ϕ_d. Some typical values are given in *Table 2.2*. In fact, band bending indicates local injection of

Table 2.2 TYPICAL VALUES OF WORK FUNCTIONS OF METALS AND DIELECTRICS AT 20°C

Dielectric	ϕ_d eV	*Metal*	ϕ_m eV
Polystyrene	4·2	Cs	1·9
Polycarbonate	4·3	Mg	3·8
Boron aluminosilicate glass	4·3	Al	4·2
Polyimide	4·4	Ag	4·3
Pure alumina (Al_2O_3)	~5	Ni	4·7
Pure silica (SiO_2)	~6	Au	4·9

electrons into the insulator; or local extraction depending on the direction of band bending. These electrons form a charge cloud analogous to the space charge around the cathode of a vacuum valve. This extends 0·01–1 μm into the dielectric from the interface.

Now for instance impurity, or radiation, can attract an electron from the valence band of an insulator. In this situation, the 'hole' left behind in an otherwise full band can itself move through the solid, rather like a bubble in a liquid. The effective charge of a hole is singly positive, i.e. the opposite of an electron. The rest of this book will discuss electron flow only but it must always be remembered that 'mirror image' hole mechanisms may occur also or instead in a given situation.

Solids conducting by means of holes (positive) are called *p*-type whereas those conducting by electrons (negative) are called *n*-type. Note that a given solid may alter from *p*- to *n*-type or vice versa as temperature, pressure, purity, etc., are altered.

We can now understand two basic types of conduction through allowed bands—by electrons through the conduction band (once they have been put there) and the 'mirror image', by holes through the valence band. In good insulators, unlike semiconductors, the electrons on the constituent ions of the lattice may bias towards such an electron or hole in transit and exert a drag. The effective mass of the electron may approach 200 times its free mass and it is then useful to consider it as a rather different entity—a 'polaron'.

The other electronic mechanism tabulated earlier was conduction through the forbidden band. This seems a contradiction in terms! Indeed, in a pure, perfect thick crystal under low fields this cannot occur.

However, if the crystal is very thin (< 3 nm) 'tunnelling' can occur. This is an effect explicable only by quantum theory which allows that there is a finite probability of an electron penetrating, rather than jumping, an energy barrier, no energy being expended.

If impurities, or structural inhomogeneities as in a glass, are present, they may cause local allowed levels ('traps') in the forbidden band. Electrons may pass through the insulator by jumping from one trap to another if they are close enough for quantum effects to

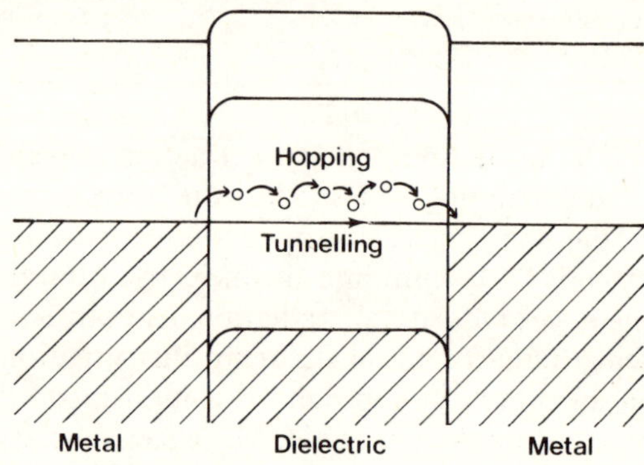

Figure 2.8. Illustration of the energy diagram showing tunnelling and hopping of electrons through an insulator. If the traps are not symmetrical across the forbidden band the Fermi level will be displaced towards the main concentration of traps

occur (< 3 nm). Simple tunnelling and 'hopping' conduction are illustrated in Figure 2.8.

It is worth noting that the width of the forbidden band of a compound tends to diminish considerably as one chooses constituent elements further down the Periodic Table. Some examples are shown in Figure 2.9. Thus heavy compounds tend to be poorer dielectrics than light ones. Indeed, heavy compounds are often semiconductors or metals. This book will therefore be dealing mainly with compounds from the first three periods of the Periodic Table. Note, however, that for these materials the forbidden band width is often large enough to be irrelevant in the understanding of

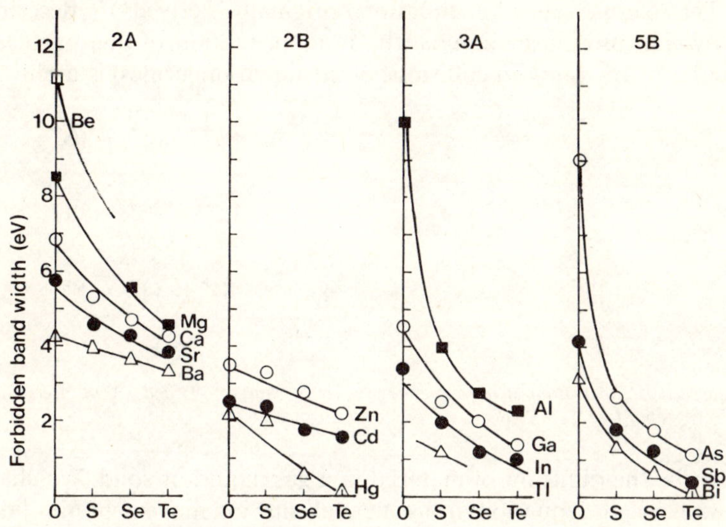

Figure 2.9. Forbidden band width of compounds with cations in indicated groups and anions from Group 6 B. [From P. J. Harrop and D. S. Campbell, Thin Solid Films, **2**, *273 (1968)]*

most electrical properties since hopping, tunnelling and ionic processes dominate.

Ionic mechanisms of conduction in solid dielectrics

There are two simple ways in which an ion may pass through the short range ordering of atoms that exists in all solids. Firstly it may move simply by squeezing through gaps in the structure, i.e. as an 'interstitial'. Secondly, it may move by virtue of one structural ion being absent, i.e. a 'vacancy' moving in the other direction. This is rather analogous to the electron hole described earlier, although here, because structural ions can be alternately positive and negative, the vacancy may be effectively negative or positive respectively. By contrast, an electron hole can only be positive. Vacancies and interstitials are illustrated in Figure 2.10. In any ionic solid there is an equilibrium concentration of vacancies and interstitials due to constituent ions jumping out of place. Impurity effects and less simple collective movements of ions will be introduced later in the book.

The Nernst–Einstein equation, originally derived to describe Brownian motion in liquids (the random motion of fine particles caused by the random collisions of the liquid molecules) is useful to

Na^+ Cl^- Na^+
Cl^- [ ] Cl^-
Na^+ Cl^- Na^+
(a)

Na^+ Cl^- Na^+
Cl^- Na^+ Cl^-
[Na^+]
Na^+ Cl^- Na^+
(b)

Figure 2.10. Illustration of (a) a vacancy (effective charge −1) and (b) an interstitial (effective charge +1) in NaCl

describe the diffusion of material in a gas, liquid or solid. We shall now derive the equation for use in analysing ion movement in solids, and particle and ion flow in liquids.

The Nernst–Einstein equation

Consider the x direction only. Assume n ions per unit volume of charge ze. Considering current flow J per unit area per unit time, from the definition of mobility μ

$$\sigma = nze\mu = \frac{J}{E} \tag{2.2}$$

$$J = nze\mu E \tag{2.3}$$

where E is the field.

At equilibrium, back diffusion will balance this drift under the field. This is one of those processes where the number passing per unit area per unit time is proportional to dn/dx. The constant of proportionality D is defined as the 'diffusion coefficient'.

Hence, from equations (2.2) and (2.3)

$$nze\mu E = zeD\frac{dn}{dx} \tag{2.4}$$

Now, since $E = -\,dV/dx$, the solution of equation (2.4) is

$$n = \text{const}\exp(-\mu V/D) \tag{2.5}$$

This must be compared with random diffusion from Maxwell–Boltzmann statistics [equation (1.35)] which gives

$$n = \text{const}\exp\{-zeV/(kT)\} \tag{2.6}$$

Consequently, from equations (2.5) and (2.6)

$$\frac{\mu}{D} = \frac{ze}{kT} \tag{2.7}$$

or, since $\sigma = ne\mu$

$$\frac{\sigma}{D} = \frac{nz^2e^2}{kT} \tag{2.8}$$

This is the *Nernst–Einstein equation*, applicable only to a perfect gas. It can be expressed more generally for liquids and solids by

$$\frac{\sigma}{D} = \frac{nz^2e^2}{fkT} \tag{2.9}$$

where σ is the conductivity caused by n similar ions of charge ze moving under low fields (concentration gradient). k is Boltzmann's constant and T the absolute temperature, f is a 'correlation factor' near to unity which allows for the subtle fact that the diffusion coefficient D relates to specific ions, whereas σ relates to defects such as vacancies. f may be calculated for a given defect in a given structure.

D is usually expressed by

$$D = D_0\exp(-Q/kT) \tag{2.10}$$

where D_0 is a constant and Q the activation energy for ion movement. Since there is ample published data for values of D in materials, the resultant ionic σ can be calculated to check whether measured conductivities are ionic or electronic. Typically D is of the order of $10^{-14}\,m^2\,s^{-1}$ for ions in glasses at 20°C.

Surface conduction

Most of our discussion is concerned with conduction and polarisation *within* the dielectrics considered. Nevertheless, with all solids and liquids there are surface conduction processes. It is convenient to define the surface conductivity of a material as the conductivity between the two opposite sides of a square of surface for fields below 10^7 V m^{-1}. Accordingly the units of surface conductivity are Ω^{-1} square. It is irrelevant what size square is taken.

2.3 INTERACTIONS OF IONS AND ELECTRONS

At this stage in the discussion the reader could very justifiably complain that the view of conduction mechanisms is inconsistent. On the one hand, the energy levels of electrons at metal–insulator interfaces are explained purely in terms of electronic charge movement to reach equilibrium. On the other hand, it is claimed that ions may move as freely as electrons in many of these materials. This would indeed be a fair criticism of much modern dielectric theory. Currently, we just do not possess the theoretical and experimental apparatus to cover the interaction of ions and electrons in practical dielectrics adequately.

Fortunately, there is some evidence emerging that the electron band pictures described earlier are valid and not commonly perturbed by ions. For instance, differences in work function quantitatively explain the charging of dielectrics by temporary contact with a metal. The reader should therefore use these concepts to describe electronic effects at those frequencies, temperatures, etc., at which they manifest themselves but recognise that the presence of mobile ions may perturb the situation.

2.4 TYPES OF POLARISING SPECIES

The term polarisation is a much abused word in chemistry and physics and it can have many meanings. In this book we use the term in one sense only: to refer to limited displacement of charge by an electrical field.

The main ways in which a substance may be polarised are illustrated in Figure 2.11. These are the mechanisms that control permittivity. Figure 2.11(a) shows biasing of orbital electrons found with all ions and atoms. This occurs extremely quickly, usually at ultra-violet frequencies.

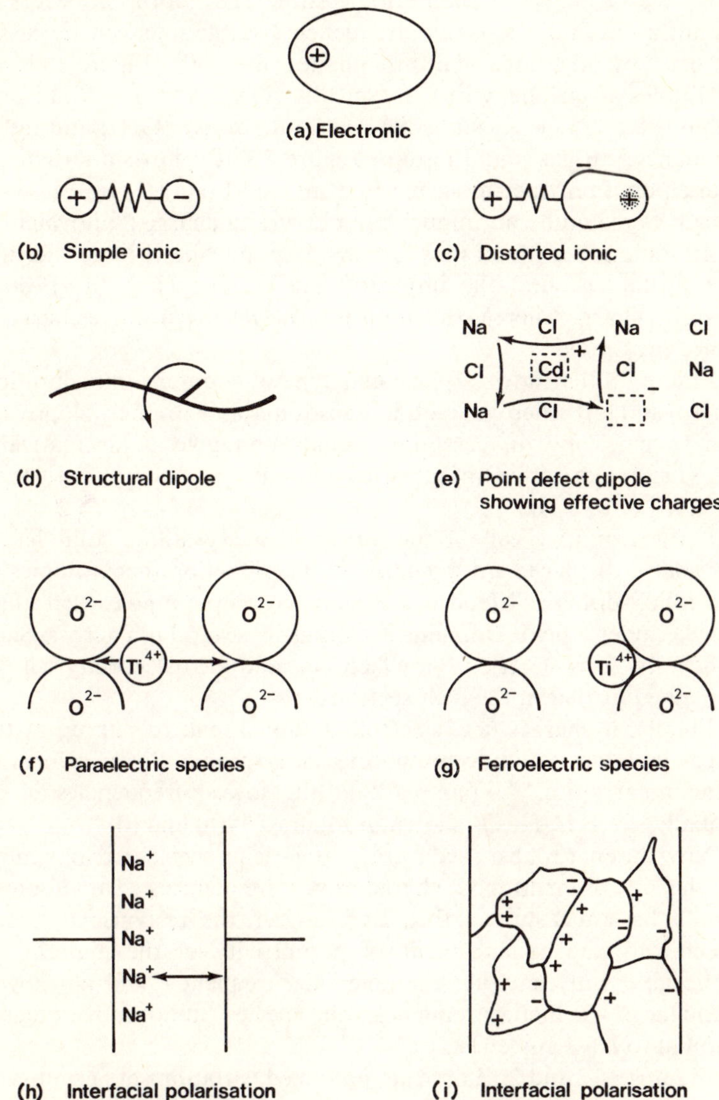

Figure 2.11. Diagrammatic illustration of various forms of electrical polarisation

Figure 2.11(b) shows simple ionic vibration where the ions can be imagined as hard spheres. This is the case with the most ionic solids, the alkali halides (metal ions from Group 1A and negative ions from Group 7B of the Periodic Table). The vibration occurs in the infra-red and microwave frequencies. Large ions can be easily distorted by other ions, and this enhances the effect [Figure 2.11(c)].

Dipoles of various types can typically vibrate at radio (MHz) and audio (kHz) frequencies in solids, and microwave (GHz) and radio frequencies in gases and liquids. Figure 2.11(d) shows a structural branch on a polymer molecule rotating and Figure 2.11(e) shows a vacancy created by an impurity ion of wrong charge ("aliovalent") in an ionic solid. Metal ions (anions) hop into the vacancy causing it to rotate around the impurity. The vacancy has an *effective* negative charge whereas the impurity has *effective* positive charge—hence the dipole.

Figure 2.11(f) shows a particular type of counter-ionic vibration [cf. (b) and (e)] where one ion in question is an extremely sloppy fit. This occurs only in crystalline solids and gives a dramatically amplified polarisability at infra-red and microwave frequencies (say, 10^9 to 2×10^{15} Hz).

If the primitive cell of the lattice in a crystalline solid has a *permanent* displacement of charge in one direction then volumes of the solid ('domains') tend to act as macroscopic dipoles that align slowly under a field. Domains are typically several μm across. Such solids are 'ferroelectric'. Paraelectricity and ferroelectricity will be explained further in the next section.

Finally, if charges (e.g. electrons or ions) tend to pile up at the edges or intercrystalline boundaries of a sample, this is a form of slow polarisation, taking place at the lowest frequencies of all (usually < 1 Hz). This is shown in Figure 2.11(h) and (i).

Now the energy absorbed by these polarising species is a maximum at the above-mentioned characteristic frequencies. Tan δ or ε'' would be a measure of this. By contrast, the response to them, specifically the enhancement of permittivity of the material, is noticeable only at the characteristic frequency and all lower frequencies. At higher frequencies the species cannot move quickly enough to have any effect.

Figures 2.12 and 2.13 indicate predicted variations of permittivity and tan δ with frequency for simple dielectrics. Regions where dielectric loss is increased and permittivity modified, as in the figures, are known as dispersion regions. At high frequencies—usually microwave and beyond—the processes that take place are undamped and are called 'resonances'. Real materials have several such resonances due to ionic and electronic polarisation.

Our main concern is with the properties of insulating materials

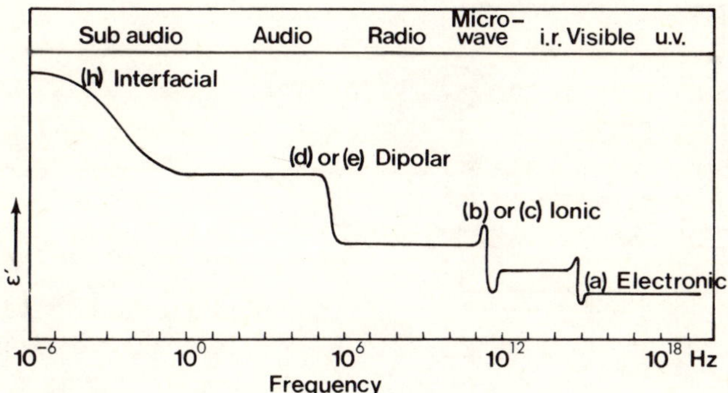

Figure 2.12. Variation of permittivity with frequency for simple dielectrics. The letters indicate the categories of Figure 2.11

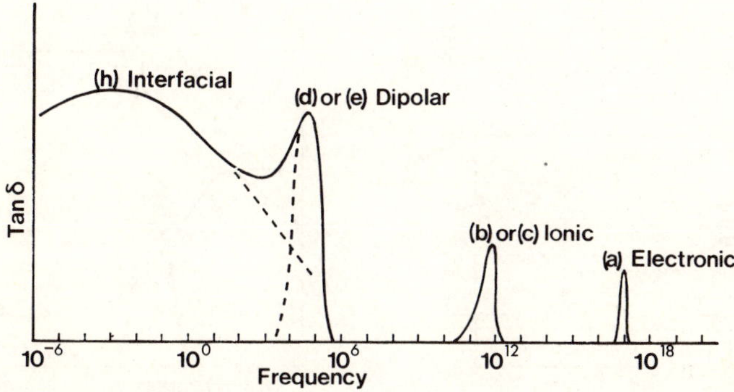

Figure 2.13. Variation of loss tangent with frequency for simple dielectrics. The letters indicate the categories of Figure 2.11

at frequencies below those of microwaves. To be slow enough to be seen here, all dispersions are heavily damped and are called 'relaxations'.

In the general case, dispersions can be simulated by the behaviour of an LCR circuit as in Figure 2.14(a). A pure resonance has no resistive terms, as in Figure 2.14(b). A relaxation has no inductive terms as in Figure 2.14(c). Clearly there can be intermediate types

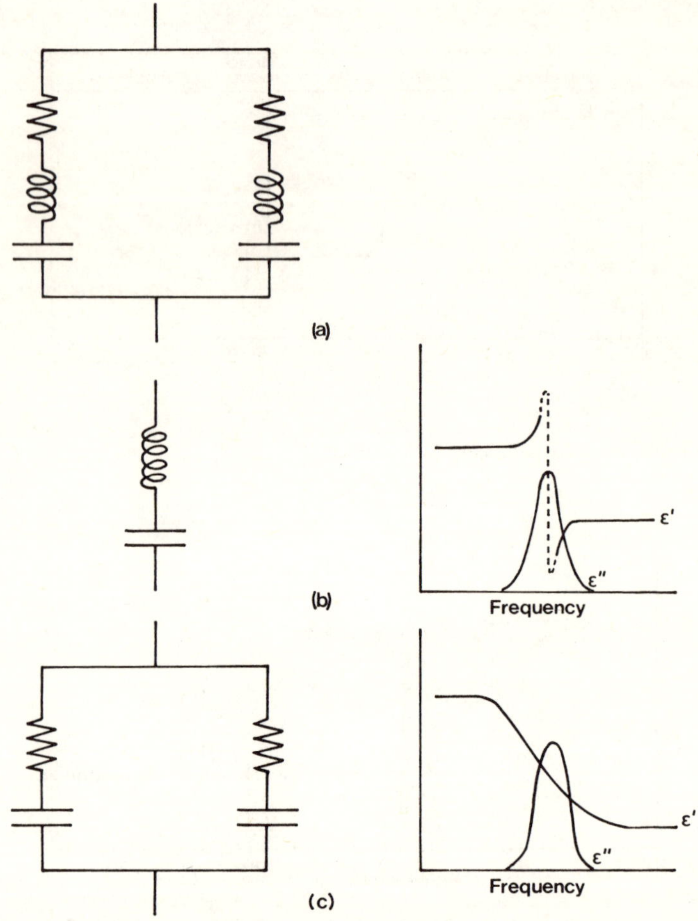

Figure 2.14. (a) General electrical analogue of a dispersion; (b) simplification to give a pure resonance; (c) simplification to give a pure relaxation. In both cases ε'' *actually reaches zero only at* $\pm\infty$

of dispersion. We shall now derive the Debye equations, which are a useful approximation for describing most relaxations.

2.4.1 DEBYE EQUATIONS

From the definition of P [see equation (1.29)] one obtains

$$P = \varepsilon_0 (\varepsilon' - 1) E \tag{2.11}$$

P and E are normally expressed as vector quantities, but we shall not need vector notation in the following argument.

Consider a *static field E*. Let bracketed suffices refer to time. The polarisation at time t, $P(t)$, has an instantaneous contribution P_1 and a time-dependent contribution $P_2(t)$.

$$P(t) = P_1 + P_2(t) \tag{2.12}$$

$P_2(t)$ increases from zero to a saturation value $P_2(\infty)$ which depends on the applied field.

Putting in boundary conditions:

(1) For $t = 0$ (i.e. $\omega \to \infty$), $\varepsilon' \to \varepsilon_\infty$, the permittivity at infinite frequency;

thus $P(0) = P_1 = \varepsilon_0(\varepsilon_\infty - 1) E.$ (2.13)

(2) For $t \to \infty$ ($\omega \to 0$), $\varepsilon' \to \varepsilon_s$, the permittivity at approaching zero frequency:

$$\text{thus } P(\infty) = P_1 + P_2(\infty) = \varepsilon_0(\varepsilon_s - 1) E. \tag{2.14}$$

From equations (2.13) and (2.14)

$$P_2(\infty) = \varepsilon_0(\varepsilon_s - \varepsilon_\infty) E \tag{2.15}$$

We now make the *main assumption* which is that $P_2(t)$ has the form

$$P_2(t) = P_2(\infty) [1 - \exp(-t/\tau)] \tag{2.16}$$

where τ is a 'time constant' or 'relaxation time'. Remembering the earlier remarks in this book that most time exponentials are solutions of an equation making rate of change of a quantity proportional to that quantity, we note that equation (2.16) can be rewritten

$$\frac{dP_2(t)}{dt} = \frac{[P_2(\infty) - P_2(t)]}{\tau} \tag{2.17}$$

Next consider a *time-dependent field*, generally expressed in complex notation

$$E(t) = E_0 \exp(j\omega\tau) \tag{2.18}$$

We can still use equation (2.16) but now $P_2(\infty)$ will be a function of time since $E(t)$ is a function of time. They are related by equation (2.15). In other words, $P_2(\infty)$ is now the saturation value of the time-dependent part of $P(t)$, i.e. $P_2(t)$, if $E(t)$ were applied instantaneously.

Therefore, using equations (2.15), (2.17) and (2.18) we have

$$\frac{\mathrm{d}P_2(t)}{\mathrm{d}t} = \frac{1}{\tau}\left[\varepsilon_0 E_0 \exp(\mathrm{j}\omega\tau)(\varepsilon_s - \varepsilon_\infty) - P_2(t)\right] \tag{2.19}$$

which solves to give

$$P_2(t) = C\exp(-t/\tau) + \frac{\varepsilon_0(\varepsilon_s - \varepsilon_\infty)E(t)}{1 + \mathrm{j}\omega\tau} \tag{2.20}$$

$C\exp(-t/\tau)$ is a transient that decays quickly with time and is of no interest.

Comparing equation (2.20) with (2.11)

$$P(t) = \varepsilon_0(\varepsilon^* - 1)E(t) \tag{2.21}$$

and in the complex notation introduced earlier in the book where $\varepsilon^* = \varepsilon' - \mathrm{j}\varepsilon''$, one has

$$\varepsilon^* = \varepsilon_\infty + \frac{\varepsilon_s - \varepsilon_\infty}{1 + \mathrm{j}\omega\tau} \tag{2.22}$$

So comparing real and imaginary parts

$$\varepsilon' = \varepsilon_\infty + \frac{(\varepsilon_s - \varepsilon_\infty)}{1 + \omega^2\tau^2} \tag{2.23}$$

and

$$\varepsilon'' = \frac{(\varepsilon_s - \varepsilon_\infty)\omega\tau}{1 + \omega^2\tau^2} \tag{2.24}$$

These are the *Debye equations*, and we find that they are reasonably applicable to most dispersions at electrical frequencies. They are therefore most important.

Of course

$$\tan\delta = \frac{\varepsilon''}{\varepsilon'} \tag{2.25}$$

so

$$\tan\delta = \frac{(\varepsilon_s - \varepsilon_\infty)\,\omega\tau}{\varepsilon_s + \varepsilon_\infty\omega^2\tau^2} \tag{2.26}$$

Sometimes, to describe peaks in ε'' (or $\tan\delta$) that are wider than the above equation implies, one substitutes $(j\omega\tau)^{1-\beta}$ for $j\omega\tau$ in equation (2.22) et seq. β is then a distribution constant making the dispersion broader and lower, and τ is the mean value of a distribution of time constants.

The Debye equations can be derived for a remarkable variety of situations. Thus a dipole rotating in a viscous medium (e.g. polar molecule in a liquid), an 'on-off' dipole rotating 180° or not at all (e.g. electron hopping between two energy wells) and a double layer dielectric all give Debye relaxations.

In the latter case, the loss peak is superimposed on a background curve, and the analysis simply consists of considering the circuit of Figure 2.15(a) as measured by the circuit of Figure 2.15(b). This is known as the Maxwell–Wagner effect.

Application of the Debye equations to given relaxations calls for a microscopic analysis of the dipole in question to give a relation

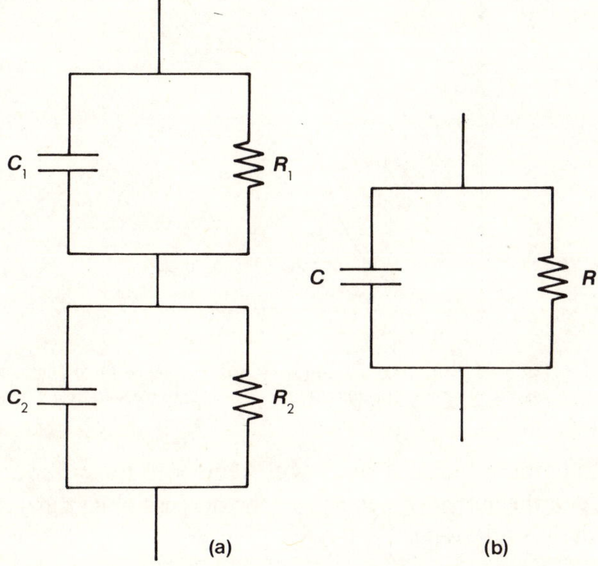

Figure 2.15. (a) Maxwell–Wagner two layer model; (b) the measured quantities

between $(\varepsilon_s - \varepsilon_\infty)$ and the number of dipoles, etc. As one might expect $(\varepsilon_s - \varepsilon_\infty)$, and hence the peak height of tan δ, is usually proportional to the number of dipoles. The anatomy of a Debye relaxation is shown in Figure 2.16. The movement of the position

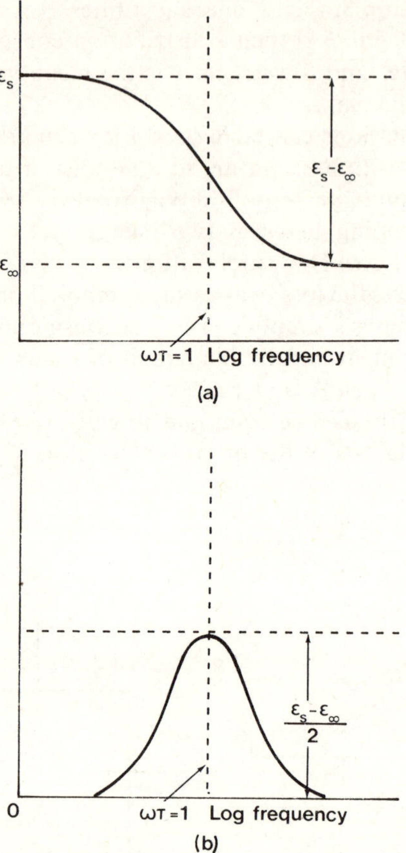

Figure 2.16. The variation of (a) ε' and (b) ε'' with frequency for a Debye relaxation. [Variation of tan δ is similar to (b) but slightly displaced in frequency.] ε'' finally reaches zero at $\pm\infty$

of the maximum in tan δ or ε'' with temperature or frequency is a measure of the activation energy of the process since τ usually obeys an Arrhenius relation

$$\tau = \tau_0 \exp(-Q/kT) \tag{2.27}$$

Note, however, that this is not true of the Maxwell–Wagner effect where the peak does not usually move.

A first necessity for the above analysis is to know what field actually acts on the dipole *in situ*. This field, E_i, is usually somewhat higher than the external field E on the sample due to the added effect of all other dipoles.

A simple case is given by the Lorentz equation

$$E_i = E\left(\frac{\varepsilon' + 2}{3}\right) \tag{2.28}$$

This assumes mutually interacting infinitely small dipoles in a cubic crystal structure. It is derived in Appendix 1. For dipoles in gases at low pressure and for small concentrations of relatively large dipoles in solids, such as point defect dipoles [Figure 2.11(e)] the dipolar interaction is small and E_i approximately equals E.

2.4.2 CLAUSIUS–MOSOTTI EQUATION

It is now convenient to derive a relation between permittivity and polarisability when the above Lorentz equation applies. We shall need this later in order to generalise about the variation of capacitance with temperature of real materials.

From equation (1.29)

$$P = \varepsilon_0 E(\varepsilon' - 1) \tag{2.29}$$

However, $P = Nm$ where m is the 'dipole moment' of the dipole and N is the number of dipoles per unit volume. (see p. 12) and

$$m = \alpha_d E_i \tag{2.30}$$

where α_d is the polarisability of a given dipole and E_i is the local field acting.

Now, from the Lorentz equation [equation (2.28)]

$$\frac{\varepsilon' - 1}{\varepsilon' + 2} = \frac{N\alpha_d}{3\varepsilon_0} \tag{2.31}$$

This is the *Clausius–Mosotti equation.*

In a later discussion it will be convenient to express it as

$$\frac{\varepsilon' - 1}{\varepsilon' + 2} = \frac{\alpha}{3V\varepsilon_0} \tag{2.32}$$

where α is the polarisability of a macroscopic volume V.

For the present, we note that this equation only applies rigorously to substances having the smallest induced dipoles, ideally those where only electronic polarisation is significant. In this case $\varepsilon' = n^2$ where n is the refractive index in a non-absorbing region. Then a derivation from the Clausius–Mosotti equation is,

$$\frac{N_0 \alpha_d}{3\varepsilon_0} = \left(\frac{n^2 - 1}{n^2 + 2}\right) \frac{M}{\rho} \tag{2.33}$$

where $N_0 = 6{\cdot}023 \times 10^{23}$ (g mol)$^{-1}$, Avogadro's number, M is the molecular weight in kg and ρ the density in kg m^{-3}.

This equation, known as the Lorentz–Lorenz equation, leads to useful insight into the electronic polarisability of non-polar polymers and liquids where $\varepsilon' \approx n^2$.

2.5 TYPES OF DIELECTRIC

Following the above sketch of the microscopic electrical properties of matter the types of dielectric can be classified. A convenient start is to take an analogy with magnetism. Simple magnetic materials are divided into diamagnetic, paramagnetic and ferromagnetic depending on whether the magnetic polarisation is created, oriented or already present. In a loose sense we take the electrical analogy by defining types as follows:

	Subdivision	*Effect of field*
Insulators (dielectrics)	Simple dielectrics	Creates dipoles
	Paraelectrics	Orients dipoles
	Ferroelectrics	Orients domains of aligned permanent dipoles

The analogy is a loose one in that whereas diamagnets and paramagnets have negative and positive magnetic susceptibility respectively, with electrical phenomena the orientation is always in the direction of the field, i.e. polarisability is always positive. We must therefore go further to distinguish the macroscopic properties of simple dielectrics and paraelectrics and we define the former as those exhibiting only the ionic and electronic polarisation of Figures 2.11(a), (b) and (c). This is characterised by a permittivity less than 40 that is little dependent on structure. By contrast, paraelectrics contain a strong permanent electric dipole in each unit cell that is always associated with permittivity greater than 20—up to 10000 or more. Those low-permittivity substances having small concentrations of permanent dipoles such as in Figures 2.11(d) and (e) are referred to as 'polar'. They are really intermediate between simple dielectrics and paraelectrics.

Ferroelectrics, as expected from the analogy with ferromagnets, exhibit polarisation that is not a linear function of applied field. The domains referred to above are regions about 1 μm across where all permanent dipoles are oriented in the same direction.

One can further subdivide ferroelectrics by analogy with ferromagnetic materials. Thus ferrielectric and antiferroelectric states exist when permanent electric dipoles in opposition respectively incompletely and completely balance out. To add to the subdivisions, the term 'electret' is used for substances that have very stable moments reminiscent of permanent magnets. That is, once polarised, they stay polarised for many years.

The above terms can be summarised as follows.

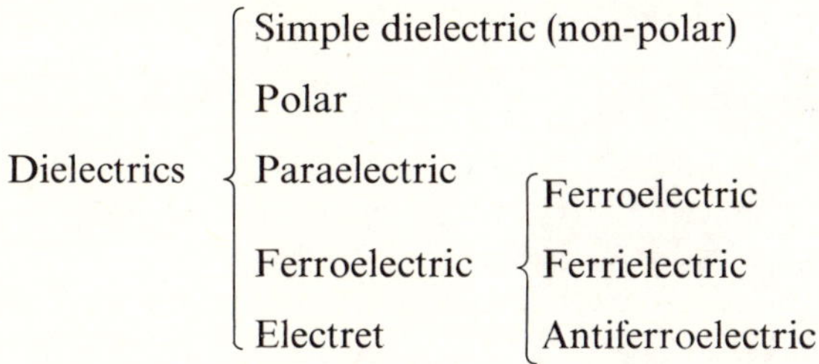

Characteristic ranges of permittivity and examples are given in *Table 2.3*.

Many substances can have different roles depending on pretreatment, temperature, etc. Thus polycarbonate is normally polar at 20°C but acts as an electret if electrically biased at an elevated temperature beforehand.

Ferro- and ferrielectric crystals have sharply temperature-dependent permittivities of the form shown in Figure 2.17. Above a characteristic temperature (the 'Curie temperature') the permittivity ceases to rise. This is due to a structural change whereby domains cease to exist, the permanent dipoles becoming merely orientable

Table 2.3 TYPICAL PERMITTIVITIES OF THE VARIOUS TYPES OF DIELECTRIC WITH EXAMPLES

Category	*Approximate permittivity range for this category*	*Examples*
Simple dielectric	1–40	Polyethylene, soda glass, vacuum
Polar gas	1–2	Water vapour
Polar liquid	4–100	Water, nitrobenzene
Polar solid	2–40	Polycarbonate
		Crystalline NaCl with Cd impurity
Paraelectric (solid)	20–20 000	Crystalline $BaZrO_3$
Ferroelectric (solid)	20–20 000	Crystalline $BaTiO_3$
Ferrielectric (solid)	1–100	Crystalline $PbZrO_3$
Antiferroelectric (solid)	1–40	Crystalline WO_3
Electret	2–40	Sulphur, $MgTiO_3$, some organic waxes

dipoles, i.e. the material becomes paraelectric. Note, however, that not all paraelectric materials become ferroelectric at a low enough temperature, and ferroelectricity is not necessarily simply related to structure.

The Curie–Weiss law is an empirical relation for the temperature dependence of permittivity above the Curie temperature whence

$$\varepsilon' = \frac{H}{T - \theta} \tag{2.34}$$

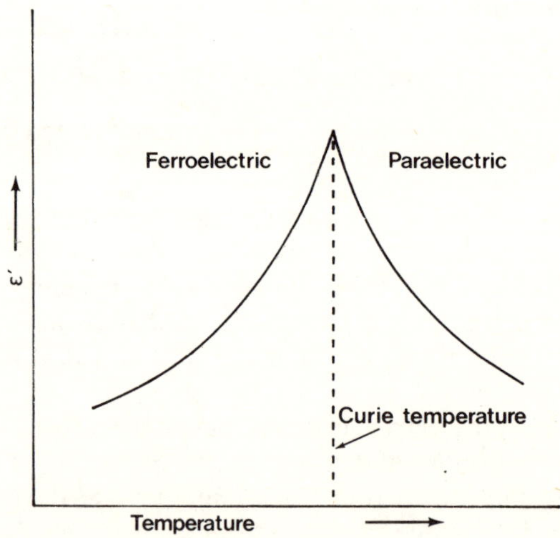

Figure 2.17. Typical variation of ε' with temperature for a ferroelectric

where H is a constant and θ is a characteristic temperature within a few degrees of the Curie temperature.

FURTHER READING

DANIEL, V. V., *Dielectric Relaxation*, Academic Press, New York (1967)

DEKKER, A. J., *Solid State Physics*, Macmillan, London (1962)

HARROP, P. J., 'Self Diffusion in Simple Oxides', *Jnl Mater. Sci.*, **3**, 206 (1968)

HILL, N. E., VAUGHAN, W. E., PRICE, A. H. and DAVIES, M., *Dielectric Properties and Molecular Behaviour*, Van Nostrand, London (1969)

ZAKY, A. A. and HAWLEY, R., *Dielectric Solids*, Routledge and Kegan Paul, London (1970)

3

Dielectrics

3.1 GENERAL BEHAVIOUR

3.1.1 TEMPERATURE COEFFICIENT OF CAPACITANCE

Although the Clausius–Mosotti equation derived earlier is strictly only applicable to dielectrics exhibiting electronic polarisation and possessing simple cubic crystal structure, like so many equations of physics it is extremely valuable as a semi-quantitative guide to the behaviour of a wide variety of materials.

The variation of capacitance with temperature is of considerable concern with many of the devices described later in this book. We recall that the temperature coefficient of capacitance, γ_c, is given by

$$\gamma_c = \frac{1}{C}\left(\frac{\partial C}{\partial T}\right)_P \tag{3.1}$$

If we take the Clausius–Mosotti equation

$$\frac{\varepsilon' - 1}{\varepsilon' + 2} = \frac{\alpha}{3V\varepsilon_0} \tag{3.2}$$

and differentiate it with respect to temperature, with some re-ordering we obtain

$$\gamma_c = \frac{(\varepsilon' - 1)(\varepsilon' + 2)}{\varepsilon}(F - \lambda) + \lambda \tag{3.3}$$

where

$$\lambda = \frac{1}{l}\left(\frac{\partial l}{\partial T}\right)_P = \frac{1}{3V}\left(\frac{\partial V}{\partial T}\right)_P$$

is the linear expansion coefficient of the material and

$$F = \frac{1}{3\alpha}\left(\frac{\partial \alpha}{\partial T}\right)_P$$

We can therefore simplify any generalisations about γ_c by plotting it as a function of permittivity for various dielectrics.

Clearly a vacuum and most gases (since their permittivity is very close to unity) must have zero γ_c. Dielectrics with low permittivity ($\varepsilon < 5$) have a temperature coefficient of capacitance that depends essentially on the nature of their bonding. Thus very ionic substances have a large proportion of their permittivity due to ionic polarisation giving a large value of F leading to a positive γ_c of a few hundred ppm per degree Celsius. Fully covalent substances having only temperature invariant electronic polarisability have $F = 0$ and

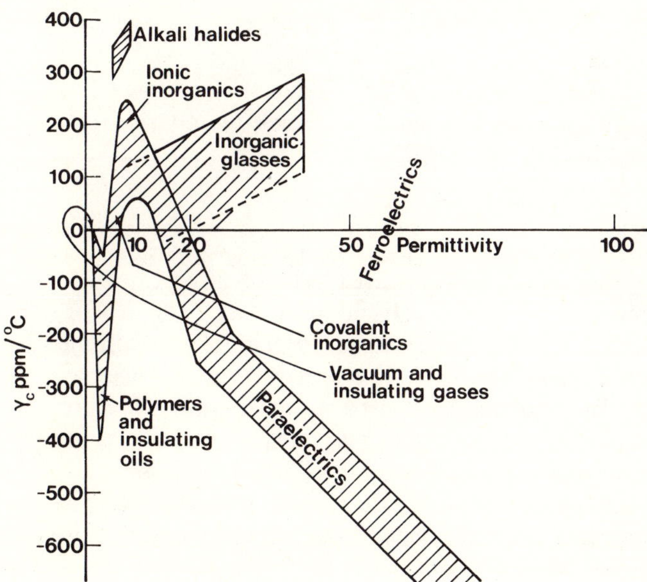

Figure 3.1. General ranges of γ_c versus permittivity for various types of solid, liquid and gas having loss tangents $\leqslant 0{\cdot}001$

hence a negative γ_c. Since the latter tend to be polymers and organic oils having permittivity near to two, γ_c is approximately equal to $-\lambda$. This is illustrated in Figure 3.1. Also shown in this figure are high permittivity glasses where the permittivity, being due to large deformable ions, has an amplified ionic polarisation term leading to positive F and γ_c.

By contrast the necessarily crystalline paraelectrics apparently have near zero F and high permittivity leading to negative γ_c given by the approximation

$$\gamma_c = -\lambda\varepsilon' \tag{3.4}$$

Of course ferroelectrics can have any value of γ_c, particularly as they can have several Curie temperatures (see Figure 2.17).

In fact the above description is over-simplified in that (*a*) it refers only to extreme types of compound, and (*b*) it implies that there are no dispersions occurring near the frequency of measurement, nor is there appreciable dielectric loss. Most of these additional phenomena would give a large added positive contribution to γ_c if $\tan\delta$ were far above 0·001.

3.1.2 D.C. CONDUCTIVITY

When a voltage is applied to a dielectric, a current passes that decays with time as in Figure 3.2. Consequently since conductivity is defined by

$$\sigma = \frac{J}{E} \tag{3.5}$$

where J is current per unit area per unit time caused by field E, the conductivity is always time-dependent. The reasons for this are many: all the polarisation mechanisms mentioned earlier contribute. However, even if one measures at effectively infinite time, one may not be observing a simple quantity since for instance polarisation at the electrodes may make the conductivity voltage-dependent and not a function of the dielectric alone.

The d.c. conductivity of a dielectric must therefore be defined with respect to time. It is usual to take the value when it is changing by less than a few per cent per hour. In the simplest cases, this may then be taken as a characteristic of the charge carriers in the dielectric moving under a d.c. field—the true d.c. conductivity.

The low field (i.e. ohmic < 10^7 V m^{-1}) d.c. conductivity of real insulating materials, particularly at the lower temperatures (<200°C), is not usually a function of the pure, perfect material, i.e.

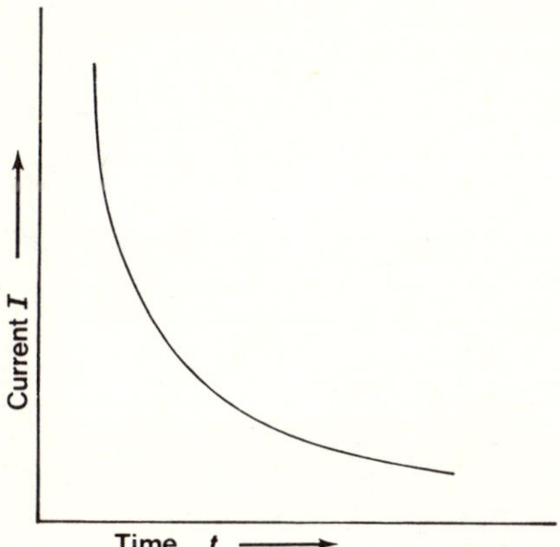

Figure 3.2. Typical current passed against time curve for a dielectric under constant voltage

it is extrinsic. It may vary radically with sample preparation and prior treatment such as heating, and may occur in localised parts of the sample rather than overall. This means that d.c. conductivity at these lower temperatures can vary by many magnitudes and not be independent of electrode area or sample thickness.

An additional complication is that when high fields are applied (>10^8 V m^{-1}) and for thin samples (say, <1 μm thick), many other conduction phenomena, particularly intrinsic ones due to electrons, can add to the 'low field' conductivity. These will be discussed in a later section on High Field Conduction.

In general the calculated d.c. conductivity, whether it be ionic or electronic, varies little with temperature at the lower temperatures but obeys an Arrhenius function over various higher temperature ranges, as processes with different activation energies dominate. Thus one typically obtains a curve similar to that in Figure 3.3. Below 20°C or so, conductivity usually becomes independent of temperature and does not obey an Arrhenius function.

To give an appreciation of the magnitude of the tiny currents that pass in dielectrics at low fields near 20°C one could say that the conductivity (in Ω^{-1} m^{-1}) of a metal is near 10^7, a semiconductor near 10 and a good dielectric near 10^{-17}. These are astronomical magnitude differences!

Generally both the mobility and the number of conducting species are activated, obeying Arrhenius equations. Since it is usually

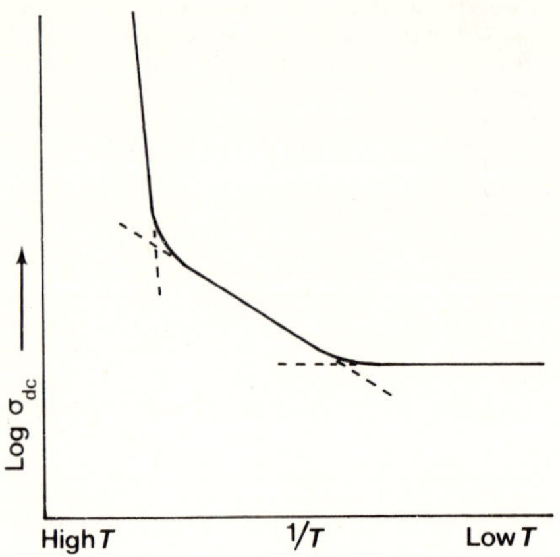

Figure 3.3. Typical variation of log σ_{dc} with $1/T$ for dielectrics

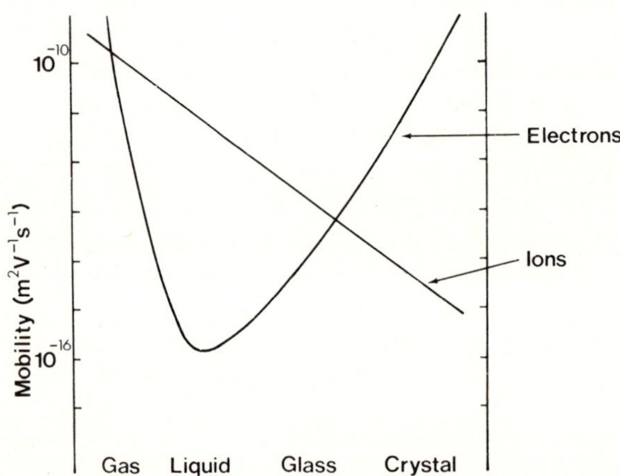

Figure 3.4. Typical variation of mobility with phase for dielectrics at 20°C

the mobility that dominates (i.e. has the highest Q) it is useful to bear in mind the general trends in the mobility of ions and electrons between the states of matter. These are shown in Figure 3.4.

3.1.3 TIME CONSTANT

Since the time constant of a material, τ, is given by $\tau = RC$ and C varies relatively little with temperature, etc., we have already noted that variation of τ is dominated by variation of R. Accordingly, the

Table 3.1 APPROXIMATE VALUES OF TIME CONSTANTS AND RESISTIVITIES AT 20°C FOR VARIOUS DIELECTRICS IN COMMON USE

Use	*Materials*	*Conductivity* ($\sigma \Omega^{-1}$ m^{-1})	*Time constant* τ *approx.*
Tuning capacitors	Polystyrene	10^{-18}	6 months
Microcircuits	Silica (SiO_2)	10^{-16}	2 days
Cables	Oil impregnated paper	10^{-15}	10 h
Transducers	Ferroelectric ceramic	10^{-12}	10 min
Electrolytic capacitors	Thin anodic oxide film	10^{-11}	100 s

above remarks concerning conductivity apply in general to the behaviour of τ. A useful guide to practical values is given in *Table 3.1*.

Time constants are sometimes quoted in ohm farads or megohm microfarads, this being exactly the same as seconds.

3.1.4 LOSS TANGENT

It was seen earlier that the loss tangent of dielectrics at electrical frequencies exhibits various types of dispersion as a function of temperature or frequency (see Figures 2.12 and 2.13). A real material may have additional processes; for instance at very low frequencies the dielectric loss will be due to the d.c. conductivity since

$$\tan \delta = \frac{1}{\omega RC} \tag{3.6}$$

usually making $\tan \delta$ proportional to ω^{-1}. At high frequencies the series resistance of the measuring electrodes R_E begins to matter. The system looks as in Figure 3.5 and as frequency is increased the dielectric conductivity is shunted by its capacitance to such an extent that the circuit appears as in Figure 3.6 giving

$$\tan \delta = \omega R_E C \tag{3.7}$$

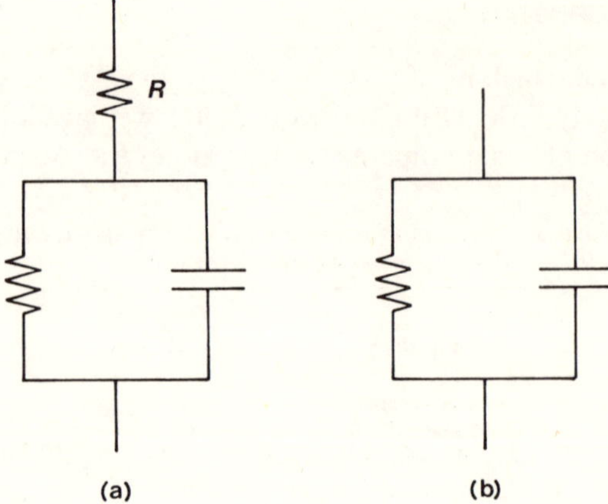

Figure 3.5. Dielectric with series resistance (a) compared with measuring circuit (b)

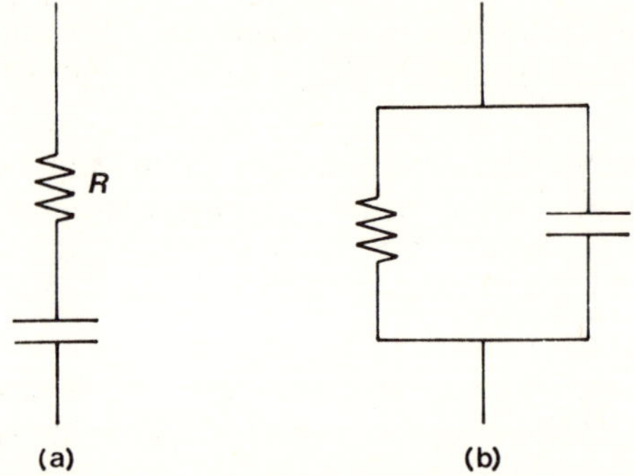

Figure 3.6. Dielectric with series resistance operating at high frequency (a) compared with measuring circuit (b)

and hence the dielectric loss becomes proportional to frequency and no longer dependent on charge carriers within the insulator.

In addition, there are often electronic and ionic dispersion processes at intermediate frequencies which can be thought of as an extremely wide distribution of activation energies giving a 'background' loss relatively independent of frequency. If we put all the above possibilities together we get a fairly general curve as in Figure 3.7(a) where the changes to be expected on increasing temperature are also shown. The concomitant variation of apparent permittivity is shown in Figure 3.7(b).

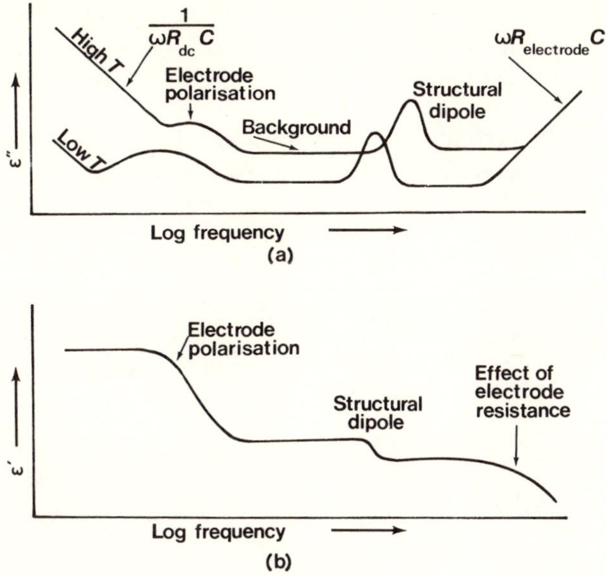

Figure 3.7. Typical variation of ε″ and ε′ with frequency for practical dielectric systems. Peaks in tan δ occur at slightly different frequencies from those in ε″ but variation is otherwise similar. In practice, electrode polarisation is sometimes negligible, and or no dipolar relaxations may be found

3.1.5 BAND THEORY FOR REAL MATERIALS

In Section 2.2 the behaviour of electrons (and their counterparts, holes) in ideal materials was discussed. One can now proceed to practical materials by generalising these thoughts a little. Firstly, recalling Figure 2.5, we acknowledge that electrons in liquids and solids no longer have the discrete energy levels they had when associated with atoms in the gas phase. These energies are broadened to bands, and insulators have the upper two allowed energy bands separated by a wide forbidden band. With liquids and glassy solids the edges of the bands are appreciably blurred and the short range order in these solids (they of course have no long range order) leads to localised allowed levels over most of the forbidden band. These localised levels can be imagined as being associated with favourable energy wells due to local disorder throughout the 'structure'. They may typically be separated by a distance of the order of 1 nm and may be drawn on the band diagram as in Figure 3.8. The same effects can be caused by localised impurities, although it is particularly easy in liquids and glasses for impurities to be neutral and thus have no electronic effects, i.e. neither (*a*) 'trapping' nor (*b*) donating electrons.

In crystalline solids the band structure is far simpler and the theory has been well elaborated in order to explain behaviour in

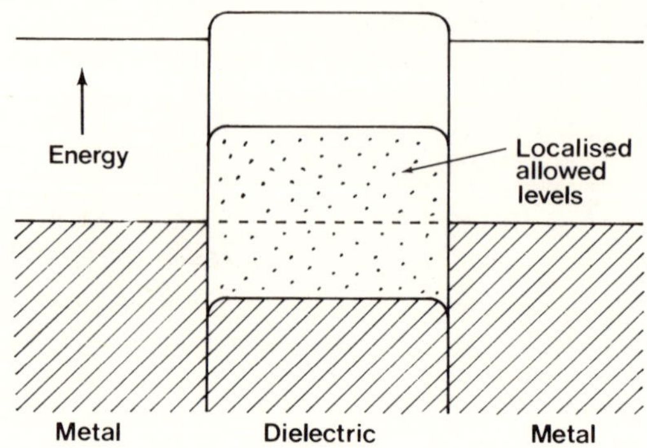

Figure 3.8. Energy diagram of a glass dielectric with metal contacts showing localised allowed levels

semiconductors. Crystals have sharp energy bands. Impurity usually causes electronic effects: it does so by creating allowed energy levels that are common to the whole solid, i.e. they are not localised in the simplest case. Electrons move freely on entering them and they

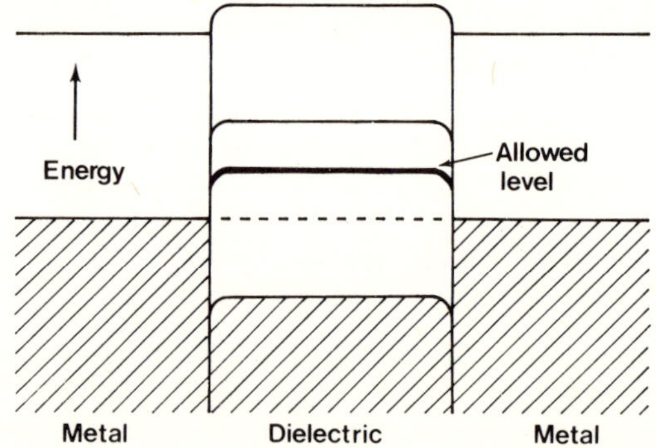

Figure 3.9. Energy diagram of a crystalline dielectric with metal contacts showing a continuous allowed level due to impurity

can be drawn as in Figure 3.9. One immediately sees, therefore, that impurity in a crystalline solid will usually cause radical changes in measured electron currents, whereas impurity in liquids or glasses rarely causes electronic effects. Of course impurity *ions* move far more easily in a glass or a liquid and marked changes in electrical conductivity can be caused by ionic conduction (cf. Figure 3.4). This is particularly true of polar materials (high permittivity), since from Coulomb's law we realise that impurity atoms can be very easily ionised because the force attracting the electron to the atom is reduced in proportion to the permittivity.

3.1.6 HIGH FIELD CONDUCTION

High fields can appreciably distort the energy bands of a solid or liquid and bring in other mechanisms of electronic and ionic

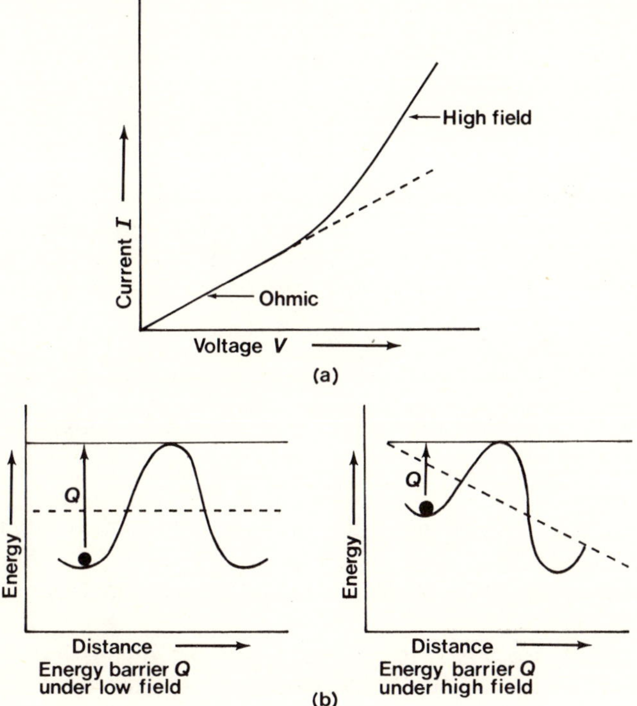

Figure 3.10. (a) Usual effect of a high field on the I/V characteristic of a dielectric and (b) lowering of an energy barrier under a high field

conduction. The effect on the I/V characteristic is shown in Figure 3.10(a). In most cases, this is due to the energy barrier that the charge carrier has to surmount being lowered by the field. The effect is shown diagrammatically in Figure 3.10(b). We summarise those mechanisms that add to the ohmic conduction already present

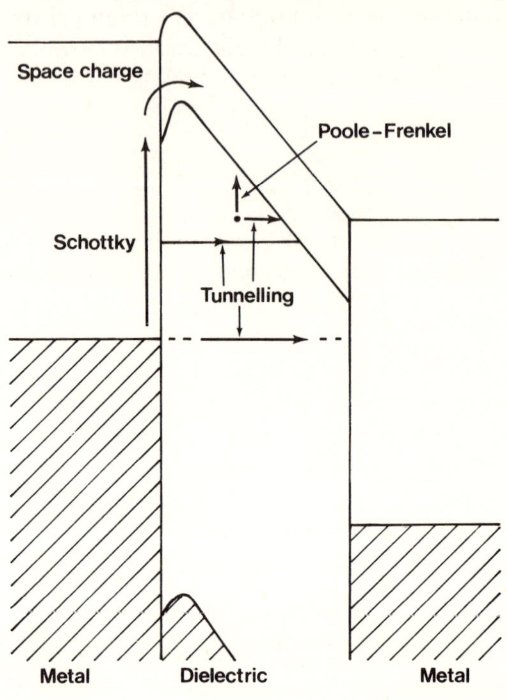

Figure 3.11. Energy diagram of a dielectric with metal electrodes subjected to a high field. This could be a thin film of alumina (Al_2O_3) with a forbidden bandwidth of 10 eV having 4 V applied

in Figure 3.11. They generally result in an increased conduction current above 10^6 V m^{-1} as in Figure 3.10. Typical dependence of the various mechanisms on voltage and temperature is summarised in *Table 3.2*.

Schottky emission is thermionic field assisted emission from the electrode by virtue of the field lowering the effective work function. It is drawn more accurately in Figure 3.16(c) later. Poole–Frenkel emission is a similar removal of electrons from local centres by virtue of the field lowering the energy barrier. Space charge limited conduction occurs when, once the electron has reached the conduction band, the band bending causes the limiting process.

Table 3.2 SIMPLEST CASES OF THE DEPENDENCE OF HIGH FIELD CONDUCTION MECHANISMS ON VOLTAGE AND TEMPERATURE

Mechanism	*Current–voltage*	*Temperature variation*
Schottky emission of electrons	$\log I \propto V^{\frac{1}{2}}$	Large, positive
Poole–Frenkel emission by electrons	$\log I \propto V^{\frac{1}{2}}$	Large, positive
Space charge limited conduction by electrons	$I \propto V^n (1 < n \leqslant 5)$	Large, positive
Tunnelling by electrons	$I = \text{const}$	Very small
Ionic, field assisted	$\log I \propto V$	Large, positive

Tunnelling is the quantum effect that allows electrons to pass adiabatically (no energy expended), provided the distance involved is well below 2 nm. Tunnelling currents are sharply dependent on the distance involved, a property exploited in strain gauges described later. Finally, field assisted movement of ions occurs when *their* energy barriers are appreciably lowered by the field. One can also have field assisted formation (dissociation) of ions in the first place.

It should be realised that high field and low field mechanisms may be basically the same. For instance ions passing over energy barriers increasingly distorted by the field give a transition from ohmic to $\log I \propto V$ behaviour. Similarly, hopping electrons can pass gradually into a non-ohmic Poole–Frenkel tunnelling regime as energy barriers are increasingly distorted.

3.1.7 ELECTRICAL BREAKDOWN

The breakdown field of a dielectric ('electric strength') is as notoriously irreproducible as the d.c. conductivity unless great care is taken. It is not only a function of frequency and temperature; it is also dependent on the nature of the electrodes, the speed with which the voltage is raised to breakdown, surface condition, mechanical stress, and softness of the dielectric. Special techniques can reduce extraneous effects with certain phases of dielectrics and these will be mentioned in the relevant sections later.

Genuinely intrinsic breakdown, a fundamental property of the pure, perfect material itself, is an elusive quantity to seek experimentally. The values of d.c. electric strength of the order of 10^7 V m^{-1} typically measured thirty years ago were at that time thought to be intrinsic. A decade later values of breakdown, at least for the best materials such as mica and one or two plastics, were found to be of the order of 10^8 V m^{-1}. These were thought to be intrinsic but in fact they were not. Nowadays the most careful experiments tend to give values around or above 10^9 V m^{-1} and the situation is further confused by the realisation that intrinsic phenomena may give lower electric strength in some cases. It would therefore be wise to treat currently reported values with reservation and to acknowledge that their interpretation is disputable.

The simplest experiments giving some of the highest breakdown strengths have been made on thin films. This is not surprising in view of the fact that with a thin film one tends to avoid particles of impurity, cracks and the like.

However, even with films there are several reasons for a reduced breakdown field E_B. For instance, electrode material can be drawn into flaws and short them out. Silver is notorious in this respect while aluminium causes little trouble probably because it easily forms an insulating oxide. This is one reason why aluminium is the favourite electrode material for capacitors and microcircuits. Thin regions and inhomogeneity can cause high local currents, concomitant joule heating and hence local melting or thermomechanical breakdown.

Three types of breakdown are common with carefully prepared films.

(1) If a low impedance (i.e. high current) supply is used, the breakdown points can travel randomly across the electrodes giving a spider-like pattern that is very much a function of the electrode used.

(2) With a high impedance supply and thick electrodes the current cannot 'run away' and the breakdown will occur by scintillation—tiny sparks at weak points of one sort or another.

(3) If the breakdown is measured away from these weak points, the breakdown may be homogeneous and it can be shown by

calculation that this breakdown may well be due to the joule heating of the solid overtaking the heat dissipation by radiation and conduction. It must be emphasised that this effect could be due to excessive ionic *or* electronic flow.

Other possible mechanisms giving the highest breakdown phenomena are electron avalanching by impact ionisation of the lattice, collective breakdown by electron–electron interaction, and 'electrochemical' breakdown due to chemical reactions resulting from excessive ionic build-up at the electrodes. In practical systems, localised spark breakdown due to some form of electron avalanching is probably most common whereas the best materials measured carefully probably exhibit *thermal* or *electrochemical* breakdown most commonly.

To some extent the proponents of the various theories test them against the observed thickness dependence of breakdown. However, the experiments made to date are really rather too variable for one to trust such an assessment and we shall therefore confine ourselves to the general trends of breakdown with thickness that are observed.

Well prepared glasses, even in bulk form, do not have gross imperfections like the grain boundaries, etc., that polycrystals have. They therefore tend to have a d.c. breakdown field that does not vary with the thickness of the sample considered. Liquids can be the same. Bulk crystals tend to have much lower breakdown strengths that increase somewhat with sample thickness, as the flaws cease to be continuous through the sample. Polymers have very high breakdown strengths that vary little with thickness.

The a.c. dielectric strength of solids and liquids is often higher than the d.c. strength provided gaseous effects are excluded. Often the generation or pre-existence of gaseous voids or bubbles in practical dielectric fluids and solids can cause a lowering of their measured a.c. electric strength. The resultant 'internal discharges' are of vital importance in practical devices and they will be dealt with in subsequent sections on gases and measuring techniques.

With homogeneous liquids and solids, one finds a high a.c. strength that increases with frequency at exceptionally high frequencies (>1 MHz) because the avalanching that tends to sparks is constantly reversed and because charged particles of all types are unable to traverse the complete distance between the electrodes before the field is reversed. The highest breakdown strength of all

E

occurs with short discrete pulses of voltage which allow plenty of heat dissipation to occur. This is such an extreme effect that even water can withstand nanosecond high voltage pulses.

3.1.8 POLYMERS

Up to now, we have tended to categorise polymers with similar phases of inorganic dielectrics. However, it would be useful to pause at this stage and consider some ways in which polymeric liquids, glasses and crystals behave uniquely electrically.

It should be noted that the relative lack of interaction between polymer chains means that electronic conduction, on the rare occasions when it occurs, tends to be along the chains. Hopping between chains is then the limiting step.

Dipolar ionic movements also have unique features, particularly associated with the fact that, with polymers, the dipoles are not independent, being attached via the main chain, nor are they small within the terms of the Lorentz equation. In particular activation energies derived from the movement of loss peaks with frequency/temperature [cf. equation (2.27)] may have less meaning due to the coupling between ions since dipole orientation is no longer random and Maxwell–Boltzmann statistics may be inappropriate. Indeed,

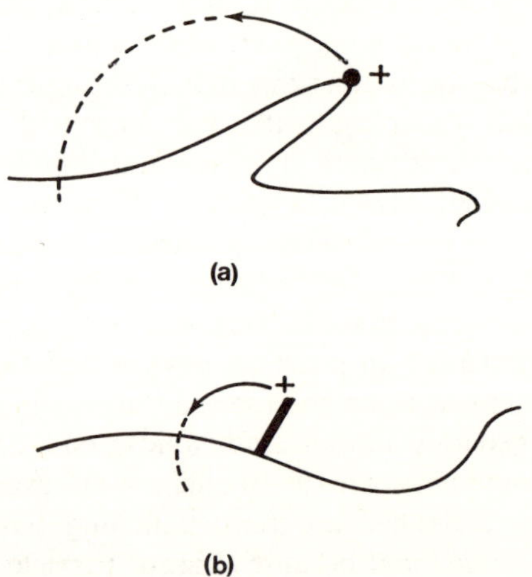

Figure 3.12. Two ways in which an impurity can form dipoles in a polymer (a) by labelling a kink and (b) by constituting a side chain

values of activation energies have frequently been quoted in the literature that are far enough above 5 eV to represent the energy necessary to vaporise the ions in question!

Since, as seen earlier, most solid polymers do not melt but go through regions of increasing softness, it is reasonable to apply the theory of liquids to such polymers. Concepts of free ions and polar side-chains moving in a viscous medium have had some success, whereas these are quite inappropriate for inorganic solids.

It is not commonly realised that solid polymers can be as impure as inorganic materials. For instance, a polymer purchased simply as 'Polystyrene' may have deliberately incorporated into it an appreciable percentage of both inorganic and organic additives identified by their function as fillers, plasticisers, nucleating agents, slip agents, etc., and all these may have electrical effects. This impurity may append itself to kinks in the polymer chain and 'label' them to form electrical dipoles as in Figure 3.12(a) or, simply, form dipolar side chains itself as in Figure 3.12(b). In either case a theoretically non-polar polymer such as polyethylene [symmetrical $(C_2H_4)_n$] may act as if it is polar, exhibiting two or more loss peaks as a function of temperature and frequency in the audio and radio frequency ranges.

Of course, it is possible that an impurity may form a counter-balancing branch that may eliminate the dipolar moment of a polar

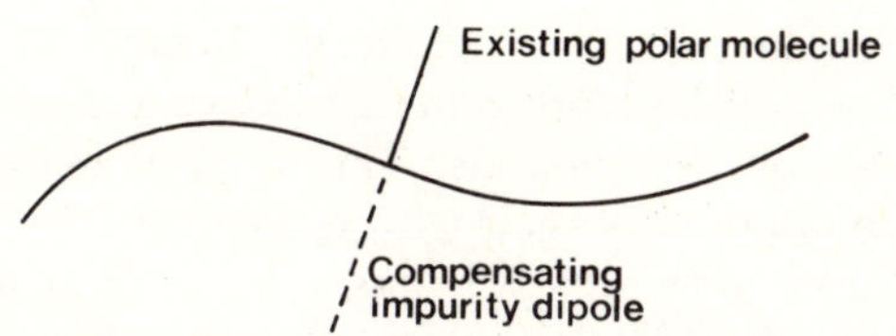

Figure 3.13. Effect of impurity in rendering a polar molecule non-polar

polymer as in Figure 3.13 thus improving the electrical properties. This usually has to be a carefully planned operation.

The breakdown of polymers tends to be more reproducible than that for inorganic materials because sharp cracks, grain boundaries and other features that increase local mechanical and electrical stress tend to be absent. However, polymers may often be so soft that they are flattened under high electric fields leading up to breakdown and this deformation must be taken into account in calculating breakdown field.

One of the most common manifestations of impurity in polymers is the motion through the polymer of free interstitial ions as with inorganic solids. However, the gaps in the structure of polymeric solids often allow molecular fragments to pass as well. This is less usual in inorganic solids.

3.1.9 EFFECTS OF WATER

It may sound rather banal to mention the effects of water in a book of this nature. However, it is extremely easy for the student to progress to practical devices or even to fundamental physics experiments and interpret them with sophisticated theories of intrinsic Debye peaks, etc., only to find later that all was lost because the many and insidious effects of water were ignored.

Water, particularly dissociated into OH^- and H^+ ions, can pass into most practical liquids and solids, and can of course exist in dielectric gases. It can enhance conductivity at microwave frequencies by virtue of its intrinsic ionic vibrations. It can enhance conductivity at various audio and radio frequencies by attaching itself to other molecules and creating large dipoles. Water can enhance conductivity below 1 Hz by causing electrode polarisation [Figure 2.11(h)].

There can of course be some significant discharge of these ions at most electrodes and therefore water can have a dominant part in the d.c. conduction of materials. This is particularly noticeable because good dielectrics, of their nature, conduct very little current. We therefore need very few OH ions, say, to upset the situation. Water in solids can create traps for electrons, thus upsetting the characteristics of the field effect transistors which will be described later.

The surface conduction of liquids and solids can also be increased by water and if surface effects are not guarded against, dipoles created by water on the surface of a given insulator can produce classical Debye loss peaks and even non-ohmic conduction. One often finds hysteresis of electrical properties as one heats and cools liquids and solids simply due to the desorption and absorption of water in a mildly humid atmosphere. One must also note that the water may be so strongly absorbed that it is not driven off until many hundreds of degrees Celsius are reached.

Needless to say the presence of traces of water can markedly reduce the breakdown strength of most materials. Further, since water is a highly polar substance it tends to encourage the ionisation of other impurities present in dielectric liquids. It is also worth noting that polymers, particularly polymeric glasses, have a remarkable propensity for absorbing water and that the structure and hence the electrical properties of the carrier material can be altered by the oxidising properties of water.

Water can be excluded by preparing samples in dry conditions and measuring them in dry conditions, e.g. in a vacuum, preferably after baking. In practical devices surface conduction is minimised by the use of glazes on ceramics, dried gases over dielectric fluids and heat sealing of polymers for example.

3.2 GASES

3.2.1 PERMITTIVITY OF GASES

Gases have low density due to the large distances between molecules. Consequently even if the molecules are polar the dipoles are usually smaller and much less prevalent than in solids and liquids, and the permittivity is still close to that of a vacuum (that is, unity). The precise value usually approximates to the square of the refractive index, i.e. it is predominantly caused by electronic polarisability. Some examples are given in *Table 3.3*.

Table 3.3 RELATIVE PERMITTIVITY OF GASES

Gas		*(Refractive index)*^{2}	ε' (20°C, 760 torr) (1 torr = 1 mm Hg)
Hydrogen	H_2	1·000070	1·000072
Oxygen	O_2	1·00054	1·00053
Argon	Ar	1·00055	1·00056
Nitrogen	N_2	1·00060	1·00060
Carbon dioxide	CO_2	1·00100	1·00096
Ethylene	C_2H_4	1·00130	1·00138

3.2.2 CONDUCTIVITY OF GASES

For low fields, gases pass incredibly low ohmic currents. For instance air typically has less than 10^{10} free ions per cubic metre, whereas even the best dielectric solid has more than 10^{11} per cubic metre. Little is known about the ohmic currents in gases since they are difficult to measure and cause little trouble in practical devices.

3.2.3 LOSS TANGENT OF GASES

Quite unlike common liquids or solids, the a.c. and d.c. low field currents that do pass in dielectric gases are generally the same. This means that their loss tangent is very low indeed, particularly at high frequencies, since it is simply given by

$$\tan \delta = \frac{1}{\omega R_{\mathrm{d.c.}} C} \qquad (3.8)$$

where $R_{\mathrm{d.c.}}$ is the d.c. resistance of the gas, a very large quantity. Thus $\tan \delta \sim 10^{-8}$ is common at 1 kHz whereas the best dielectric solids and liquids tend to have $10^{-5} < \tan \delta < 10^{-2}$.

3.2.4 BREAKDOWN OF GASES

Since gases can have such low d.c. conductivities and loss tangents they have many uses as simple insulation. Their exceptionally low permittivity and very poor electric strength at common pressures limits their use in capacitors. Nevertheless under certain conditions, gases have some of the highest electric strengths. For instance, for typical electrode separation, the electric strength of dry air at atmospheric pressure is only around $2 \times 10^{7}\ \mathrm{V\,m^{-1}}$ whereas a high vacuum ($\sim 10^{-5}$ torr where 1 torr = 1 mm Hg) has ten times this electric strength, comparable to that of the better solids. It therefore becomes clear that the main uses of dielectric gases are in high voltage insulation and this explains why by far the most work has been concentrated on breakdown.

We have to be careful in our definition of breakdown with a gas as there are far more modes than with liquids or solids. Also, the nomenclature is very often wrongly applied. Let us consider breakdown as occurring when the gas becomes markedly conducting.

Uniform field breakdown of gases

A simple beginning will be to examine the current–voltage characteristic for an increasing uniform d.c. field applied to cold parallel plate electrodes. Initially one obtains a low field ohmic conductivity, due to electron emission from the electrodes, and ions and electrons created in the gas by cosmic rays, light, etc.

Eventually this saturates (because all charge carriers created are collected), goes through a transition region (as secondary processes provide more carriers), and finally, at the highest fields breakdown occurs. This is illustrated in Figure 3.14. 'Photoionisation' is

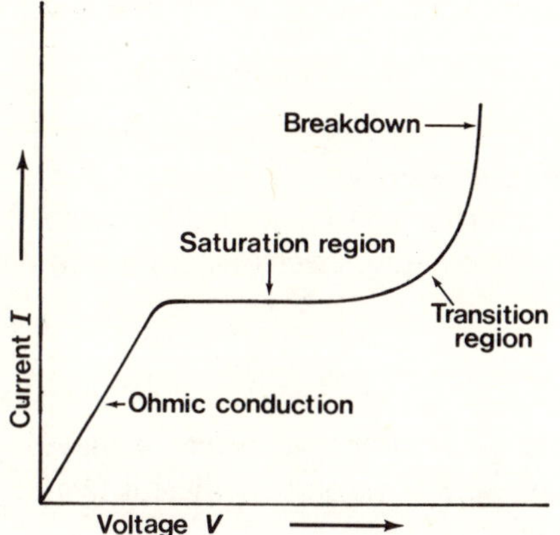

Figure 3.14. Typical V/I plot for a gas subjected to a d.c. potential. Breakdown is usually by the Townsend mechanism giving a glow discharge or a spark, provided currents are limited

the main cause of the transition region. This is a process by which ultra-violet light produced by high energy collisions in the gas excites electrons from the cathode. The voltages and currents involved and the mode of breakdown depend on the gases involved but also depend radically on electrode separation and gas pressure, and the current supplied by the external circuit.

Paschen plot

To initiate and maintain uniform field d.c. breakdown between cold electrodes in a gas, two quite distinct processes must take place: (*a*) any electrons moving out from the cathode must multiply by impact ionisation of the gas; (*b*) a steady stream of initiating electrons must be created at the cathode—this is usually done by collision ionisation of the metal by incoming ions.

Townsend showed that the multiplication obeyed the expression

$$\frac{dn_x}{dx} = \alpha n_x \tag{3.9}$$

where n_x is the number of electrons passing through unit area per unit time at distance x from the cathode. α, Townsend's first coefficient, is a constant of proportionality equal to the average number of ionisations per unit distance of travel by one electron. Hence

$$I = I_0 \exp(\alpha d) \tag{3.10}$$

where I is the current passing due to an initial flux I_0 of cathodic electrons, d being the electrode separation.

Townsend further showed that, for maintenance of the breakdown, one needs a process creating electrons at the cathode so that

$$I = I_0 \frac{\exp(\alpha d)}{1 - \gamma\,[\exp(\alpha d) - 1]} \tag{3.11}$$

at initiation of the glow discharge, where I_0 may be due to one electron and γ is Townsend's second coefficient relating to the production of cathodic electrons.

This leads to a breakdown condition

$$\exp(\alpha d) - 1 \approx \exp(\alpha d) = 1 \tag{3.12}$$

which becomes meaningful when we note that detailed theory and experiment shows both γ and αd to be functions of V/Pd where V is the breakdown voltage and P is the pressure. It is therefore clear for a given system that V is a function of Pd alone, a fact first observed experimentally by Paschen. Typical Paschen plots are shown in Figure 3.15. At high pressures the breakdown is localised in the form of a spark. At pressures below atmospheric the breakdown is usually a non-localised glow known as a 'glow discharge'.

It should particularly be noted that the breakdown strength of a gas between atmospheric pressure and a moderate vacuum is very poor. A millimetre gap can only support a few kilovolts whereas

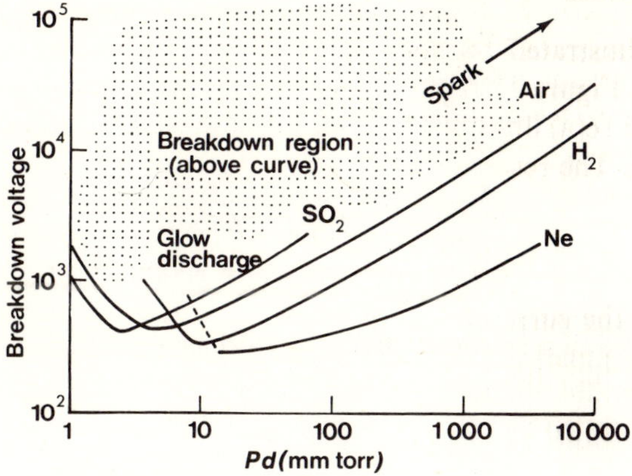

Figure 3.15. Paschen plot for various gases

most solid and liquid insulators support 0·1–10 MV. This is why gas voids in cables are their weakest feature.

At pressures greater than a good vacuum, for a given electrode separation, increase in pressure makes it very much more difficult for breakdown to occur. This is why high voltage X-ray sets used for therapy and diagnosis in hospitals have been made much smaller in recent years by the utilisation of high pressure gaseous insulation.

V is high at high pressures and separations because the mean free path of charge carriers is low. V is high at lowest pressures and separations because there are too few ionisable molecules.

At pressures above atmospheric, spark breakdown may occasionally originate from more localised effects than those described by Townsend. Two mechanisms known as *kanal* and *streamer* have been put forward but due to their limited field of application they will not be elaborated here.

Enhanced emission from an electrode of low work function

It is now useful to consider four special cases of conduction—not breakdown—in gaseous dielectrics that have practical importance in vacuum tubes. They refer to artificially enhanced electron emission from an electrode of low work function. For simplicity, they are illustrated by band diagrams for a metal electrode in a vacuum in Figure 3.16.

Figure 3.16(a) illustrates thermionic electron emission enhanced by heating. The resulting current, I, obeys an Arrhenius-like equation

$$I = AT^2 \exp(-\phi/kT) \tag{3.13}$$

where I is the current per unit area, A a geometrical constant and ϕ the work function of the electrode.

This is known as the *Richardson–Dushman equation.* In practice,

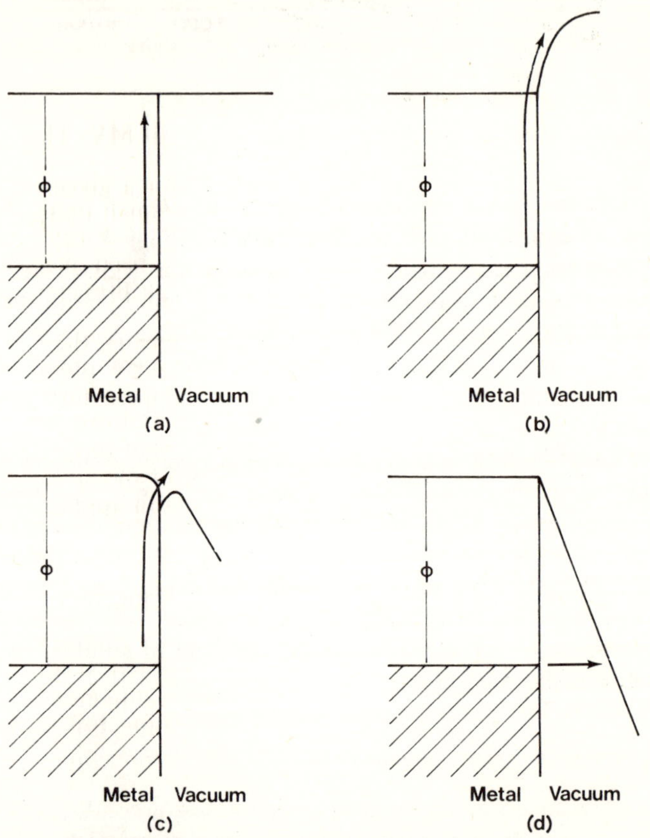

Figure 3.16. Four modes of enhanced electron emission from a metal electrode into a vacuum: (a) thermionic emission; (b) thermionic emission impeded by a space charge, i.e. space charge limited conduction; (c) Schottky emission; (d) tunnelling

thermionic emission is only appreciable for electrode coatings of very low work function such as heavy oxides, held at temperatures in the region of 1000°C.

Figures 3.16(b), (c) and (d) refer to mechanisms that have already been illustrated for metal–dielectric contacts in Figure 3.11. Figure 3.16(b) shows *space charge limited conduction.* In the present context we can consider this to be due to thermionic emission providing electrons faster than they can be removed. Figure 3.16(c) shows a reduction of barrier height by field, i.e. *Schottky emission.*

In the present case one can substitute $\phi - kV^{\frac{1}{2}}$ for ϕ in equation (3.13) to describe this, where k is a constant. (This gives $\log I \propto V^{\frac{1}{2}}$ as in *Table 3.2* earlier.)

Figure 3.16(d) shows the effect of extremely high fields in making the barrier thin enough for tunnelling to occur. For the simple triangular barrier shown, this is known as *Fowler Nordheim tunnelling*, where

$$I = BV^2 \exp(-\beta/V) \qquad (3.14)$$

B and β are constants containing the work function. The reason Fowler Nordheim tunnelling gives a more complex relation than $I = \text{const}$, as quoted for simple tunnelling in *Table 3.2*, is that the tunnelling distance is itself a function of field in this case.

Arcs

Spark breakdown can sometimes cause sufficient local heating of an electrode that excessive thermionic emission occurs. 'Arcs', another mode of breakdown, may then result. This is particularly encouraged when high currents can be supplied by the external circuit. Arcs are self-sustaining bright tubes of ionisation stretching from cathode to anode and maintained by thermionic emission at the former. They are exploited in arc lamps.

Corona

With non-uniform fields or a.c., glow discharges and high pressure sparks are initiated at peak voltages of the same order as those for d.c., but prior effects often mask this and very different and earlier breakdown is usual.

Fields changing quickly with distance or time trigger various additional types of avalanching of current carriers and the resulting light emission and localised breakdown is known as corona. Corona depends radically on temperature, electrode separation and geometry, and the nature and pressure of the gas. Typical appearance of glow discharge, spark, arc and corona breakdown are compared in Figure 3.17. Corona usually has a tree-like appearance

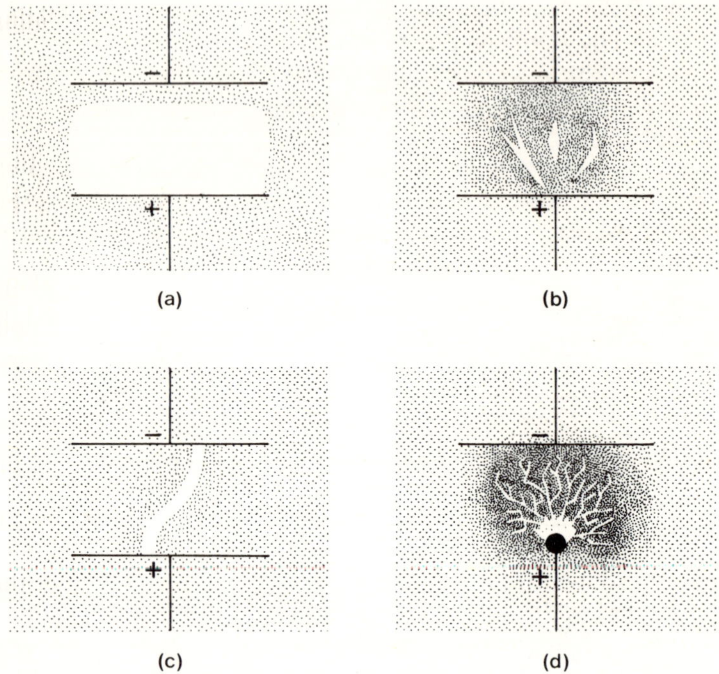

Figure 3.17. Typical appearance of (a) glow discharge, (b) spark, (c) arc and (d) corona breakdown in a gas

radiating from an intensely glowing region as shown. Sometimes only a localised glow is observed. Corona usually follows the trend of the Paschen plot but at lower voltages. Accordingly corona breakdown is also difficult at highest or lowest pressures.

Corona may be intermittent or continuous. Principal theories acknowledge that the earlier-mentioned Townsend avalanching can occur in the glow and branching of breakdown. However, the high speed of propagation additionally calls for heavy photoionisation ahead of the tree and large local enhancement of the electric field by the ionic space charge at the tips of the tree.

Suppression of breakdown in gases

Since all modes of breakdown in gases rely on a supply of electrons, any method that restricts or diverts this supply will tend to suppress breakdown. Use of inert gases is beneficial as they are not readily ionised. However, electrons may still be supplied from electrodes in this case. More desirable is the presence of atoms that readily 'mop up' electrons in the creation of negative ions. The negative ions tend to be large and immobile and the attached electrons are therefore effectively prevented from contributing to breakdown. Compounds containing fluorine and chlorine are particularly noteworthy in this respect and the practical uses of gases such as sulphur hexafluoride (SF_6) will be discussed later.

Nomenclature

The term streamer is sometimes applied to all localised modes of breakdown. The term corona is frequently and wrongly applied to breakdown in general: the term 'discharge' is to be preferred here. The term 'spark' is sometimes applied to a Townsend event whether localised or not.

3.3 LIQUIDS

3.3.1 PERMITTIVITY AND D.C. CONDUCTIVITY OF LIQUIDS

Dielectric liquids used in practical systems tend to consist of covalently bonded molecules that are relatively non-polar. This is because ionic compounds in the liquid form have high conductivity due to the free movement of the constituent ions, whereas at the other extreme highly polar covalently bonded molecules permit ready ionisation of the slightest impurity by virtue of their high permittivity (cf. Coulomb's law) and the permittivity of the water that they tend to attract. It is convenient to classify liquids into four types in order of increasing permittivity as in *Table 3.4*.

Clearly the Type 4 (strongly polar associated) liquids have little potential in practical insulation. They have a small application in ultra-fast pulse shaping devices where their high permittivity can be exploited provided the pulses are exceptionally short (nanoseconds)

Table 3.4 TYPES OF LIQUID DIELECTRIC

Type	*Examples*	*Permittivity approx.*	*Behaviour*	*D.C. Conductivity* $\Omega^{-1}\,m^{-1}$ *when purified*
1. Non-polar	Transformer oil	2	Poorly ionising	10^{-14}
2. Mildly polar	Chlorinated diphenyl (e.g. Aroclor in capacitors)	6	Poorly ionising	$\leqslant 10^{-8}$
3. Strongly polar but not associated	Nitrobenzene Propylene carbonate	$\geqslant 20$	Fairly poorly ionising	10^{-12}–10^{-10}
4. Strongly polar and associated	Water, Ethanol, Hydrogen cyanide	$\geqslant 20$	Very ionising and often self-dissociated	$\geqslant 10^{-8}$

and infrequent. The ions then have too little time in which to move and propagate extensive spark breakdown.

Type 3 (strongly polar, not associated) liquids should have several electrostatic applications in energy conversion and storage once the formidable purification problems are mastered. In the meantime only Types 1 and 2 find extensive practical use.

With compounds of Type 1 and 2, impurity ionic and charged macroscopic particle d.c. conduction dominate, since, as we noted, electrons find it difficult to hop between molecules. Indeed, in many practical systems it is possible to explain low field and high field d.c. conduction and even breakdown in terms of the movement of charged dust, etc., between the electrodes. Conduction currents are so small that they can often be ascribed to a few tens of specks of dust of the order of 1 μm or less in diameter.

Walden's rule

In the presence of appreciable liquid flow (e.g. for small currents) the diffusion of particles and ions (the usual charge carriers) can obey the Nernst–Einstein equation derived earlier

$$\frac{\sigma}{D} \approx \frac{nz^2e^2}{kT} \tag{3.15}$$

where σ is conductivity, D diffusion coefficient, n number of carriers of charge ze. These charge carriers also approximate to spheres moving in a viscous medium, obeying the classical Stokes–Einstein relation

$$\eta r D \propto kT \tag{3.16}$$

where η is viscosity and r radius of sphere.

Since $\sigma = nze\mu$ where μ is mobility, combining the above two equations gives Walden's rule

$$\eta\mu \propto \frac{1}{r} \approx \text{constant} \tag{3.17}$$

for a given charge carrier in a given liquid independent of number, temperature and pressure. Walden's rule even has some limited application to ionic movement in solid polymers where we have already observed that mechanical properties can have much in common with liquids. Further sophistication with liquids includes consideration of local order.

3.3.2 LOSS TANGENT OF LIQUIDS

Since, in preparing and modifying dielectric fluids for given applications, we must be so concerned to reduce ionic and particulate movement, it is useful to have a simple means of detecting these carriers. Of course they can be detected by the interfacial polarisation shown in Figure 2.11(h) manifested by a very low frequency dielectric loss peak for low applied fields as drawn in Figure 2.13. However, as we shall learn in a subsequent chapter on the measurement of dielectrics, it is rather awkward to measure dielectric loss peaks at very low frequencies and although such peaks may be moved into a higher frequency range by increase in temperature they would then tend to be swamped by d.c. conduction [see Figure 3.7(a)].

Garton effect

It is more convenient to measure low frequency polarisation by speeding up the ions and/or particles in their path between the electrodes not by temperature but by high fields. One can then measure a peak in dielectric loss as a function of applied a.c. field at an easily measured frequency, say 50 Hz. This is known as the Garton effect and is illustrated in Figure 3.18. The peaks can also

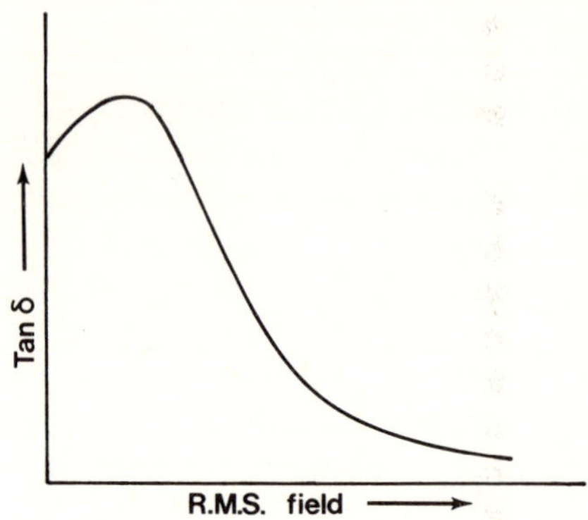

Figure 3.18. Garton effect characteristic. A broad peak indicates a wide variety of charge carriers

be characteristic of ions or particles moving shorter distances, say between the interstices of an oil-filled paper. Here one would expect that the higher the frequency of the peak the smaller the characteristic distance of relaxation of the ion, and indeed one can analyse dielectric loss peaks as a function of low frequency, and the Garton effect, in terms of the type of charge carrier and its distance and ease of movement.

High fields speed up ions and particles by quite different mechanisms. the energy barriers are distorted within the liquid giving the field-assisted conduction of *Table 3.2*. By contrast, particles can not move faster in transit but they are charged to higher values when making contact with interfaces, permitting conduction to increase by increments between each traverse.

Dielectric absorption of liquids

A more practical effect that is primarily due to the interfacial polarisation that causes low frequency loss and the Garton effect is 'dielectric absorption'. When a given dielectric has a field put across it, and this field is then removed, the electrodes are briefly shorted out (not enough for all the ions and particles to relapse back) and the device left on open circuit one finds subsequently that it has partially charged up again.

This is due to the slow-moving charge carriers remaining in position, as shown in Figure 3.19. It can be a frightening phenomenon

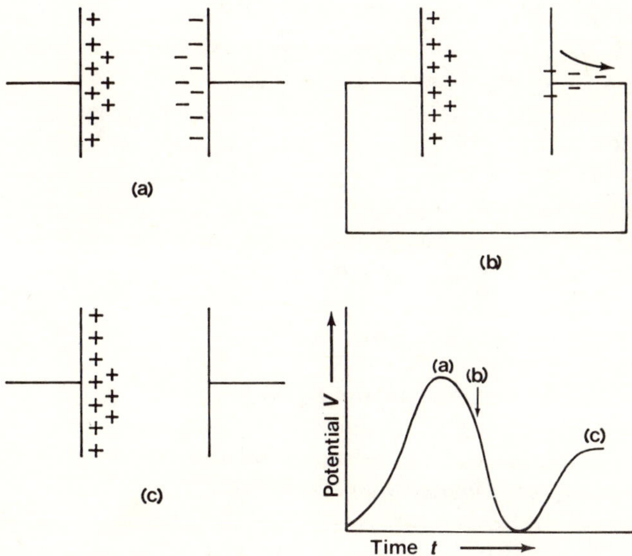

Figure 3.19. Schematic drawings showing (a) capacitor charged, (b) briefly discharged with slow-moving positive charge remaining, and (c) partially recharged on open circuit

with practical dielectrics since a person who has briefly discharged a charged insulator may subsequently touch it and be severely shocked. It can also be a nuisance with small capacitors in analogue computers, in that the charge on a capacitor is used as a reference point, and drift of potential after shorting represents error.

Dielectric absorption is therefore something to beware of in practice. Oil-impregnated constructions and solid ceramics are particularly prone to the effect. Polystyrene is particularly free of it. It is not a particularly good analytical tool for the characterisation of charge movement if only because it can be partly due to dipoles and inhomogeneities.

3.3.3 PRACTICAL CONSIDERATIONS FOR LIQUIDS

Dielectric liquids tend to react with adjacent materials such as the metal parts of a transformer or a cable. The products of the reaction usually create conducting species. One is also concerned about the higher vapour pressure and flow properties of liquids as opposed to solids and their degradation as a consequence of over heating. Hence for a given application of a dielectric fluid one not only compromises electrical properties with cost and ease of handling; one is also vitally concerned with compatibility, viscosity, sensitivity to contamination (i.e. selecting as low a permittivity as possible, consistent with other properties) and working temperature range.

There must be a final warning on charge transport in liquids. The particles concerned usually cause movement of the surrounding liquid. Consequently the shape of the container, electrodes, etc., must be closely considered in any full analysis.

3.3.4 BREAKDOWN OF LIQUIDS

Just as the a.c. conductivity and the low frequency d.c. conductivity of the best dielectric fluids is often a function of particulate impurity, so the breakdown of these materials is also radically dependent on extrinsic factors. Macroscopic charge carriers—dust perhaps—can act as 'stress raisers' rather like lightning conductors and this can trigger breakdown by virtue of the high local electrical stress causing avalanche creation of other charge carriers and consequent over-heating or sparks.

It is also true that compatibility problems become the more acute at the high stresses leading up to breakdown. The nature of the electrodes and of any adjacent gaseous atmosphere can be a controlling factor for instance. Electrodes should be smooth and chemically inert.

When a practical dielectric fluid breaks down it does so in three stages of increasing stress, possibly due to particulate contamination. These are reminiscent of boiling:

(1) noisy bubbling at the electrodes;
(2) the production of non-collapsing bubbles which pass through the fluid; then
(3) actual breakdown, usually with associated local heating and chemical degradation of both electrodes and the liquid in question.

Generally the d.c. electric strength of a dielectric fluid varies little with thickness. However, it tends to be time-dependent, since bubbles may grow only slowly.

Inhibition of breakdown in liquids follows the same lines as inhibition of breakdown in gases. Compounds that are not easily ionised are useful (e.g. Type 1 of *Table 3.4*). Even better are compounds that absorb electrons to form relatively immobile negative ions. These generally contain fluorine or chlorine atoms. Ability to immobilise any small positive ions evolved is also desirable.

Design of a high voltage dielectric fluid centres on sensitivity to salt and aqueous contamination during manipulation, and heat dissipation. The effects of breakdown products caused by minor discharges in triggering chemical deterioration or runaway currents are important. For instance, castor oil is often used in pulse transformers and discharge capacitors despite its low permittivity because it breaks down to form large molecules rather than reactive gases, and absorbs any hydrogen evolved.

3.4 GLASSES

3.4.1 PRODUCTION OF GLASSES

Some compounds form glasses far more easily than others. It is difficult to generalise about the ease with which organic materials form disordered structures, since the small intermolecular forces make it relatively easy to prepare most of them as glasses simply by fast cooling.

By contrast inorganic compounds generally do not form glasses at all easily. Ionic compounds in particular have very large interionic coulombic forces making them form the simplest crystalline structures with great agility. Thus the alkali halides, the epitome of ionic solids, are virtually impossible to prepare as glasses at all.

Zachariasen criteria

Zachariasen long ago proposed simple rules for glass formation. Considering the glass as a three-dimensional network lacking long-range periodicity, he required that for a glass to be easily formed one must have:

(1) no oxygen atom linked to more than two metal atoms;
(2) a minimum number of oxygen atoms surrounding a given metal atom;
(3) oxygen polyhedra sharing corners with each other rather than edges or faces and forming three-dimensional networks; and
(4) at least three corners of each oxygen polyhedron must be shared.

An illustration in two dimensions is shown in Figure 3.20. The Zachariasen criteria imply that oxides having a metal ion M in a molecule M_2O or MO cannot form glasses, whereas M_2O_3, MO_2 and M_2O_5 can. Complex glasses are formed by non-glass-forming cations resting in the large interstices within the latter three types of compound.

This gives a basis for rationalising a large number of common inorganic glasses: it does not account for the large number of anomalous bulk glasses. Indeed these criteria are quite inappropriate when we consider thin inorganic films prepared by modern vacuum and electrochemical techniques for we find that amorphous inorganic *films* can be readily prepared from almost any compound except the alkali halides.

The traditional glass-forming molecule is silica (SiO_2) which forms a basic structure consisting of a tetrahedron of three oxygens with silicon over them. This short-range order is distributed randomly through the solid. As expected from the Zachariasen criteria, Al_2O_3 and Ta_2O_5 form excellent (thin film) glasses whereas MgO reverts to cubic crystals.

Mixed alkali anomaly

In order to manipulate and shape silica-based glass dielectrics, one usually adds alkali metal ions (Li, Na, K) which locate themselves in holes in the structure and lower the softening point. Unfortunately, they can also migrate through the solid under electric fields.

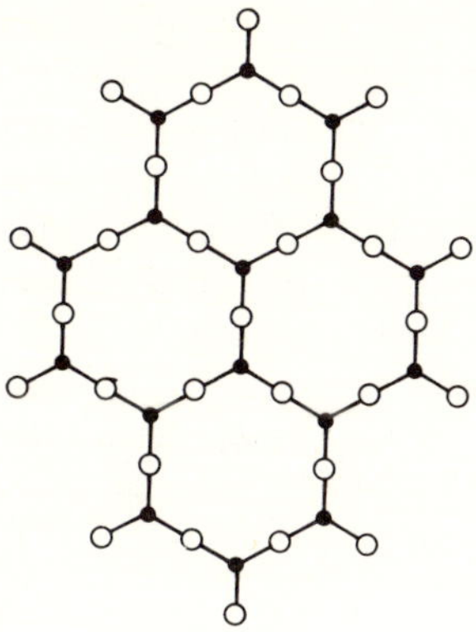

(a)

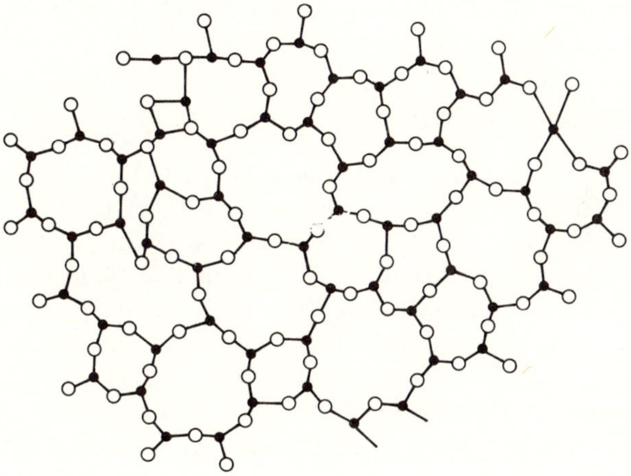

(b)

Figure 3.20. Two-dimensional representation of the structure of (a) a crystal lattice, and (b) a glass showing bonds between constituent atoms

The electrical effects are not additive since the lattice relaxes around the alkali metals and the presence of one metal ion hinders movement of another for large concentrations. This is called the Mixed Alkali Anomaly.

In practice, however, one can only produce the best glass dielectrics by either (*a*) excluding alkali metals entirely or (*b*) by blocking their motion with large heavy ions such as Pb and Ba. Thus pure silica layers have many electrical uses in microcircuitry and one of the best substrate materials for miniature electrical circuits is Corning 7059, a barium aluminosilicate glass. In the latter case, the glass must be produced in the form of extremely flat, clean sheets and pure silica could not be manipulated. It must also be resistant to subsequent adventitious contamination such as from sodium in the salt on the human hand. Alkali metal ions placed on silica glass move freely through it, by means of the spaces in the structure mentioned earlier.

3.4.2 SURFACE CONDUCTION ON GLASSES

The above criteria do not apply to surface properties. The surface conductivity of a glass particularly depends on its affinity for moisture and is a function of its chemical nature.

3.4.3 MECHANICAL CONSIDERATIONS FOR GLASSES

Whilst glasses have excellent compressive strengths, their tensile strengths are poor and they usually have little resistance to thermal shock. Nevertheless when their dimensions are small, many mechanical failure mechanisms cannot occur. Thus thin films and the best glass ceramics, which consist of specks of crystalline phase separated by a fine lacework of glass phase, have excellent thermal and mechanical properties. Unfortunately glass ceramics usually call for the addition of nucleating agents that somewhat degrade their electrical properties and limit their usefulness as dielectrics, and they are as yet rather expensive.

3.4.4 PERMITTIVITY OF GLASSES

Given that under favourable conditions glasses of most inorganic compounds can be prepared it is useful to get a general guide as to the permittivities available. Of course with a glass one cannot have permanent dipoles such as those causing paraelectricity and ferroelectricity or indeed polar materials in the normal sense of the word: the necessary crystallinity is not there. Consequently glasses have relatively low permittivity. The lowest permittivity is exhibited by amorphous polymers because their ions are so small and covalently bonded that their permittivity is due to electronic polarisation alone, roughly equalling the square of the refractive index.

As one takes inorganic glasses and increases the proportion of deformable ions—these are the large ones—one can enhance the permittivity since the charge on one ion deforms the other and this 'self-polarisation' strengthens the bond and hence polarisability and permittivity [see Figure 2.11(c)]. However, one eventually comes to the stage when insertion of ever larger and more deformable ions in the structure results in a disproportionately smaller *number N* of polarising species per unit volume and this outweighs the benefits. This can be seen from the Clausius–Mosotti equation

$$\frac{\varepsilon' - 1}{\varepsilon' + 2} = \frac{N\alpha_d}{3\varepsilon_0} \tag{3.18}$$

Consequently we can increase the polarisability of the dipole α_d so long as we do not decrease N disproportionately. In practice one finds that the maximum permittivity of inorganic glasses is in the

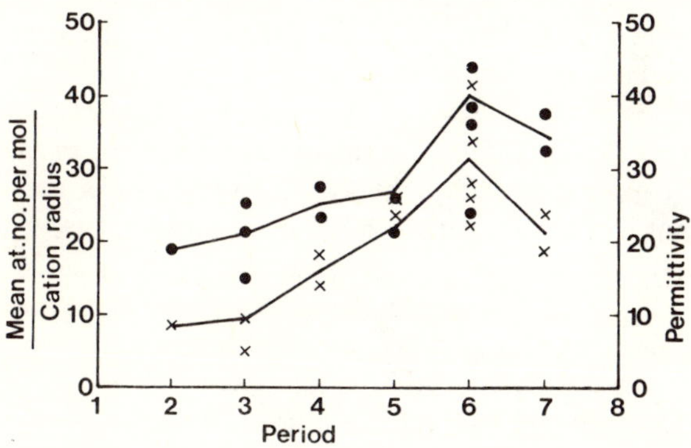

Figure 3.21. Permittivity (×) and mean atomic number/cation radius (●) versus cation period for oxide glasses. [After P. J. Harrop, Jnl Mater. Sci., 4, 370 (1969)]

region of 40, by virtue of incorporating an optimum amount of Period 6 cations. This is illustrated for oxides in Figure 3.21. Of course one cannot obtain good insulators by having both heavy cations and anions because this would be associated with very small forbidden band gaps (see Figure 2.9 and associated discussion) and the resulting electronic conduction mechanisms made possible would probably make them conductive semiconductors—outside the scope of this book.

3.4.5 EFFECT OF IMPURITY IN GLASSES

Glasses of their nature have many desirable properties that make them valuable in a wide variety of dielectric applications. It has been noted that they have good compressive strength. Their relatively low permittivity implies that they do not ionise impurities at all easily and indeed electrons created by any such ionisation cannot move freely since they tend to be trapped by structural defects. Therefore, in practice, one finds that a given glass may contain several percent of various impurities and yet show no resultant electrical effects. By contrast a crystalline inorganic solid—like conventional semiconductors—may show changes in conductivity of many magnitudes as a consequence of the incorporation of a few parts per million of impurity.

The greatest electrical weakness of glasses is that they, rather like liquids, tend to have large interstices permitting any small charged impurity ions that are present to pass through relatively freely. We have noted that this is true of Na^+ and K^+. Two other important examples are Ag^+ and Cu^+ also from Group 1 of the Periodic Table. Thus with ordinary window glass, which contains a lot of sodium, one can put a potential across at a few hundred degrees Celsius and readily plate out sodium on one electrode. In electrical devices, one usually avoids copper or silver electrodes, because they donate ions.

With glasses then, provided one particularly avoids the presence of Group 1 cations, either within the material or within the electrodes, one can produce superb insulation that is remarkably insensitive to most other forms of contamination. As noted, one can even do something about the small ions mentioned above by plugging the holes in the structure with large ions such as barium.

3.4.6 LOSS TANGENT OF GLASSES

In 1957 Stevels produced the generalised loss spectrum shown in Figure 3.22 that has since been employed by a large number of glass technologists. In fact Stevels' spectrum is a considerable simplifica-

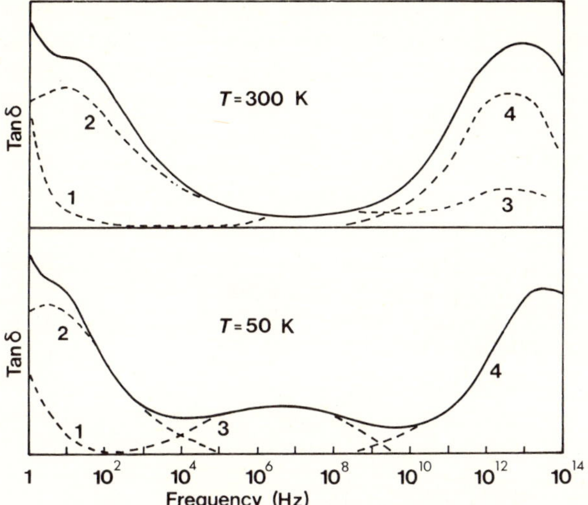

Figure 3.22. Stevels' generalised spectrum of losses for glass illustrating also the effect of temperature. [From J. M. Stevels, 'The Electrical Properties of Glass', *Handbuch der Physik*, **20**, 350 (1957) by permission of the author]

tion of the general behaviour outlined earlier in this book. He claims that there are four primary mechanisms in order of increasing frequency.

(1) d.c. conduction loss where $\tan \delta = 1/\omega R_{d.c.} C$.

(2) Ionic relaxation loss. Specifically, this refers to ions migrating limited distances in the glass and not to rotation of dipoles. Since there is a broad distribution of structural energy barriers in a glass it is expected that this peak will be wider than a Debye peak. Since it is so loosely defined it may be taken to apply to our earlier electrode polarisation peak at ultra-low frequencies *or* short paths in the structure. The important experimental observation is that loss below 1 kHz tends to correlate with d.c. conductivity and suggest an allied mechanism.

(3) Deformation loss. This is taken to be due to structural ionic dipoles (within the short-range order) such as Si–O bonds rotating. It is only seen at electrical frequencies at low temperatures.

(4) Vibration loss. Stevels includes this for completeness, but of course these are the intrinsic vibrations of the ions themselves, centred on infra-red frequencies. These are resonances.

Although Stevels' spectrum is a useful basis in understanding bulk glass at least, it should be realised that it tends to ignore any possible electronic contributions. It is also true that for most glasses one should substitute for peak (2) an ultra-low frequency peak due to electrode polarisation followed by frequency-independent loss tangent [cf. Figure 3.7(a)].

In the 1940s Garton, Gevers and Du Pré were aware of the region of frequency-independent loss and produced generalised equations to describe it. One particular success of this treatment was its prediction of an addition to the temperature coefficient of capacitance γ_c calculated from the Clausius–Mosotti equation in Section 3.1. The additional term, due to conducting species hopping over a wide distribution of energy barriers, approximated to $0{\cdot}05 \tan \delta$. (Note that in Section 3.1 this was avoided by specifying materials of low loss.)

3.4.7 BREAKDOWN OF GLASSES

With glasses, as with other materials, one can reduce thermal effects giving breakdown by using pulses of short duration or holding the sample at a low temperature. This is particularly important with inorganic glass as it has such poor resistance to thermal shock. In this event, specimen thickness and the nature of the electrodes matter little. Where thermal effects matter, breakdown strength decreases with increase in specimen thickness, temperature and voltage duration.

Understandably, in the thermal régime, electrode materials can be important (e.g. Ag ions may diffuse into the glass) and low alkali, low conductivity glasses have the highest breakdown strength. Thermal breakdown has been calculated taking the high field current as obeying the empirical relation $I = A \exp BE$ which we realise from *Table 3.2* is probably due to field-assisted ionic conduction, where A and B are constants. Prestressing with d.c. causing build-up of space charge, can reduce impulse breakdown strength.

3.5 CRYSTALLINE SOLIDS

3.5.1 DESIRABILITY

It has been seen from the preceding discussion that organic insulators conduct primarily by the rotation of polar radicals and the movement of extrinsic ions or electrons. Accordingly, there is no reason to believe that the crystalline phase of a given polymer is necessarily more insulating than the amorphous phase. Each case must be taken on its merits.

Set against this we have learned that amorphous inorganic materials are highly desirable if only because they are more insensitive to impurity than their crystalline counterparts. It would appear from this that one would be constantly avoiding crystalline inorganic materials ('ceramics'), but in fact there are several specific benefits that make them desirable in specific applications.

For instance, for working at high temperatures one requires materials with high melting point and large forbidden band gaps since forbidden band gaps tend to decrease linearly with increasing temperature. These materials tend to be the lightest inorganic ones. However, this calls for simple compounds chosen from the early periods of the Periodic Table such as MgO, which, as expected from the Zachariasen criteria, cannot be easily made into a bulk glass. Indeed it often means that one must employ an ionic material and tolerate its polycrystalline form from the electrical viewpoint. Actually polycrystals often have better resistance to thermal shock —a vital consideration in spark plug insulation for example, and they do not soften at high temperatures.

Some examples of crystalline simple dielectrics in practical use are given in *Table 3.5*. Note that they all consist of elements from the first three periods of the Periodic Table.

Polar polymers may have dipole rotation inhibited on passing from an amorphous to a crystalline phase. A tendency towards crystallinity in a polymer can also give enhanced mechanical strength and ready cleavage or extrusion for fibres. Paper [primarily crystalline cellulose $(C_6H_{10}O_5)_n$] is a porous material useful as an insulating support for dielectric fluids. OH side groups make cellulose slightly polar, giving $\varepsilon' \sim 6{\cdot}5$ at the expense of $\tan \delta \sim 0{\cdot}005$.

Rubbers, having a spring-like ordered structure, are broadly a form of crystalline dielectric, their value lying in their shock absorbent and sealing properties. Natural rubber has the formula $(C_5H_8)_n$. A common artificial rubber, butadiene $(C_4H_6)_n$, has a very similar spring-like molecule. Rubbers can have radial side arms that cancel out, having no electrical effect. However, tangential groups may have an additive effect giving high permittivity at certain frequencies

Table 3.5 TYPICAL PROPERTIES AND USES FOR CRYSTALLINE SIMPLE DIELECTRICS

Name	*D.C. conductivity* 20°C, $\Omega^{-1}\,m^{-1}$ *within a few magnitudes of*	*Permittivity*	*Loss tangent* 20°C, 1 kHz	*Upper working temp* °C	*Advantages*	*Disadvantages*	*Typical use*
Porcelain (basically aluminium silicate)	10^{-14}	6	0·005	150	Will form large complex shapes with low thermal expansion, strong, high electric strength and resistance, unaffected by moisture.	Rather conductive at high frequency and temperature	Insulators under electric railway track and for suspending overhead power cables
Mullite (basically aluminium silicate)	10^{-10}	8	0·005	1800	High working temperature	Less easily shaped	Electrically wound furnace tubes
Steatite (basically magnesium silicate)	10^{-10}	7	0·005	1000	Reasonable resistivity at high temperature and frequency	Less easily shaped	RF coil formers, variable capacitor insulation

Corundum (basically aluminium oxide)	10^{-12}	7	0·002	1800	Strong; high temperature operation Fairly high permittivity yet low loss above 1 GHz	Fairly expensive	Spark plug insulation Strip lines (miniature microwave components for e.g. radar)
PTFE [polytetrafluoroethylene, $(C_2F_4)_n$]	10^{-18}	2	0·00005	280	Excellent electrical properties at high frequencies. Easily shaped; strong	Fairly expensive, poor corona resistance	Electrometer insulation, wire sleeving
Mica-single crystal (basically aluminium silicate)	10^{-13}	6	0·001	300	Easily cleaved to 50 μm platelets. Excellent corona resistance	Properties vary from batch to batch	Insulation of motor commutators
Sapphire and spinel single crystals (basically aluminium oxide and magnesium aluminate respectively)	10^{-18}	7–8	0·0001	$\geqslant 200$	Suitable crystal structure for oriented growth of silicon overlayer, excellent electrical properties	Expensive	Substrates for certain field effect transistors

coupled with excessive loss making them undesirable for insulation. This is illustrated in Figure 3.23. A rubber may be cross-linked to harden it up a little for some applications without unduly degrading

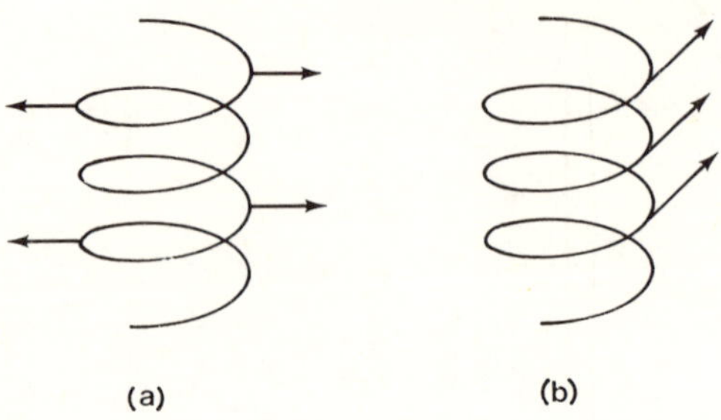

Figure 3.23. Polar elastomers with (a) compensating radial side groups and (b) additive tangential side groups

the insulation. Vulcanisation consists of cross-linking with sulphur.

An example of a rigorous requirement for an insulating rubber is in connectors for electrical circuitry in supersonic aircraft, where operation temperatures may approach 200°C and inertness to hydraulic fluids and other contamination is important. Fluorinated silicone rubbers are appropriate.

3.5.2 PARAELECTRICS AND FERROELECTRICS

One may continue with these examples but of course a comprehensive list would be very long indeed. It is more profitable to continue the discussion with those crystalline materials that are used as dielectrics for the far simpler reason that they have outstanding *electrical* properties caused by crystallinity. These are primarily the paraelectrics and various types of ferroelectric. Examples of the latter are given in *Table 3.6*. Unlike simple dielectrics, ferroelectrics in common use often incorporate the heavier cations, down to Period 6, in the Periodic Table.

Ferroelectrics have been briefly discussed in Section 2.5. They exhibit hysteresis of polarisation P as a function of applied field E as shown in Figure 3.24. This is caused by growth of the domains (regions of aligned permanent dipoles) oriented in the field direction due to an adjustment of crystalline structure and is therefore relatively slow. It can often be watched by shining polarised light on the specimen.

Initially, at A, the domains cancel each other out. If after reaching B (where the process is approaching saturation) the field is reduced, the polarisation lags until it eventually begins to saturate with a large enough negative field. On going positive the 'lag' is repeated. Note, therefore, that the permittivity of a ferroelectric must be quoted for a given field strength, history and temperature, since it varies sharply with all three.

Table 3.6 EXAMPLES OF FERROELECTRIC MATERIALS

Crystal		*Curie temp.* (°C)	*Effect associated with*
Potassium dihydrogen phosphate	KH_2PO_4	−151	H bond
Lead titanate	$PbTiO_3$	487	Ti ion
Barium titanate	$BaTiO_3$	10 and 120	Ti ion
Cadmium titanate	$CdTiO_3$	−210	Ti ion
Lead niobate	$PbNb_2O_6$	570	Nb ion

It was earlier noted that the permittivity of a ferroelectric rises sharply with temperature (Figure 2.17). This is because increased thermal energy encourages the growth of oriented domains. Then, at the Curie temperature, the domains vanish as the thermal energy jumbles up the individual dipoles and the substance becomes paraelectric. The permittivity of the paraelectric then proceeds to *drop* with further increase in temperature because the primitive dipoles tend to become increasingly randomly oriented by thermal agitation, despite the aligning effect of the field.

Ferroelectricity is most commonly and dramatically seen in inorganic crystals where a necessary requirement is the absence of a centre of symmetry in the crystal structure. However, although one can postulate various other structural conditions for the phenomenon to occur, no reliable predictions can yet be made as to what materials will or will not be ferroelectric at some temperature. Recently, slight discontinuities in permittivity v. temperature reminiscent of ferroelectricity have been seen with polymers and genuine polymeric ferroelectrics may well be developed.

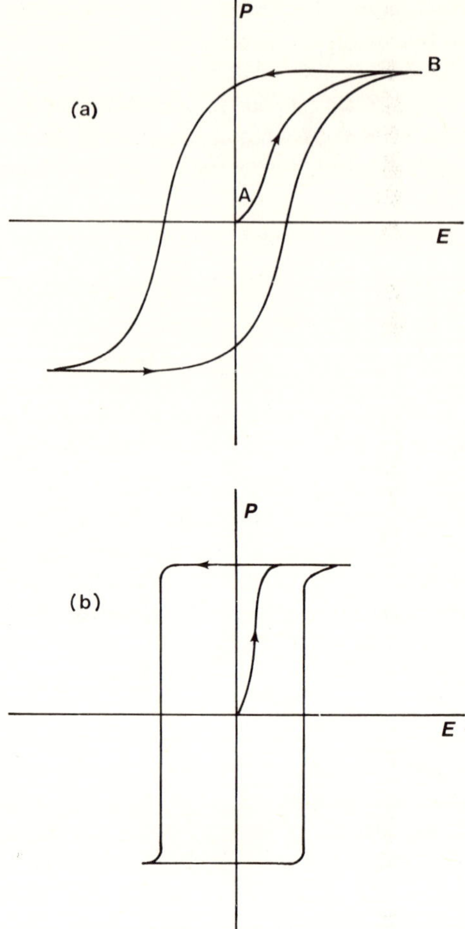

Figure 3.24. Ferroelectric hysteresis (a) for a polycrystal; (b) for a single domain crystal

Barium titanate has been more closely studied than any other ferroelectric and the structural causes of its behaviour are quite well understood. Above the main Curie temperature (120°C), the atomic structure is symmetrical, of the type known as cubic. It consists of octahedra with oxygen ions (of charge −2e) at their corners and titanium ions (of charge +4e) at their centres. The barium ions are uniformly distributed in the larger interstices between octahedra. The substance is paraelectric in this state.

Below the Curie temperature a structural transition occurs by which the titanium ions are all displaced towards corresponding oxygen ions in their surrounding octahedra. This large displacement of charge causes ferroelectric behaviour as it can be reversed by the field.

There are, in fact, various more complex features of the electrical behaviour of barium titanate but these can generally be related to crystal structure.

Some ferroelectrics only polarise along one crystalline axis. They are known as 'uniaxial ferroelectrics'. Others polarise along several axes and are 'multiaxial ferroelectrics'.

Curie temperatures may vary widely. Some materials exhibit 'blurred' transitions (useful in capacitors that must have a high and steady capacitance) and multiple transitions. Unfortunately ferroelectrics or paraelectrics with high permittivity tend to have high loss tangent as expected from relaxation theory. Certain ferroelectrics, such as those based on Pb—La—Zr—Ti, turn coloured or opaque on application of an electric field.

3.5.3 PIEZOELECTRICS

The domains in most ferroelectrics may be permanently aligned as in Figure 3.25(a) by applying an electrical stress at paraelectric temperatures and cooling. Mechanical shearing stress as in Figure 3.25(b) then causes charge displacement, known as piezoelectricity.

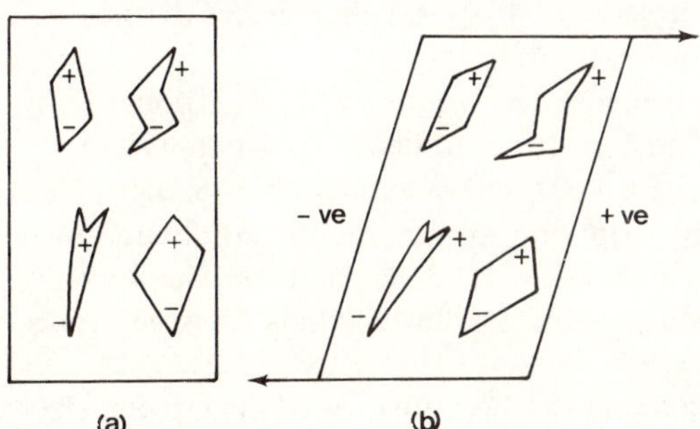

Figure 3.25. (a) A ferroelectric with oriented domains. (b) Mechanical stressing causing charge displacement

Piezoelectricity can be monitored by electrodes placed on either side of the specimen. The device then acts as a transducer, converting mechanical into electrical energy or vice versa. If a.c. electrical fields of the right frequency are applied, ultrasonic waves may be transmitted into the environment.

Piezoelectricity is not confined to ferroelectric materials. It is found with most crystals having an anisotropic structure. For example single crystals of SiO_2, known as quartz, can be cut so as to give electrical pulses on deformation. Quartz is a simple dielectric and therefore has no domains; there is no need for prior electrical biasing for piezoelectric effects to be seen in this case. Rolled polyvinylchloride film is piezoelectric.

3.5.4 ELECTRETS

In principle, any dielectric can contain electric charges that can at a given temperature be oriented by an electrical field. If the substance is then cooled so that the charges are immobilised, it will act as the electrical equivalent of a permanent magnet. It will be an electret.

In fact, electrets produced in this way are called thermoelectrets. Similar devices can be produced by creating asymmetric bound charge with radiation but this is a special case, being intermediate between the volume polarisation of a true electret and the surface charge on a biased dielectric (capacitor).

Just as a magnet produces a static magnetic field, an electret produces a static electric field. The charge species causing the electret phenomenon can be ions or charged molecules piling up at electrodes or intercrystalline electrodes. Alternatively, it can be oriented molecular dipoles, point defect dipoles or even 'hopping' electrons or electron holes.

An electret loses its volume charge exponentially with time but the time constant (time to decay by 1/exp) is usually of the order of 100 years. By contrast, a capacitor is a dielectric that is surface charged by applying an e.m.f. with no change in temperature. As discussed in Section 3.1.3, capacitors usually have much shorter time constants—from a few seconds for some types to six months for polystyrene.

Materials that can be transformed into strong electrets are usually at least partly crystalline. Examples include polycarbonate, various fluorocarbons, magnesium titanate and forms of crystalline sulphur.

It is rather surprising that electrets still tend to be omitted from introductory textbooks as the term 'electret' was first used by Oliver Heaviside in the latter part of the last century when he was one of the first people to postulate the existence of the phenomenon. Eguchi and others gave actual demonstrations from 1920 onwards using Carnauba wax and beeswax thermoelectrets.

In the last ten years, polyurethane, polyethylene, polycarbonate and other electrets have been studied, particularly with a view to producing thin films of electret for various applications.

By strict analogy with ferromagnetic materials in magnetism, an oriented conventional ferroelectric such as barium titanate would be generically an electret. However, ferroelectrics usually have time constants of hours only. Although they have an electric dipole in every crystalline unit cell this merely enhances permittivity. It does not produce a large and stable static electric field. Other flaws in the analogy with magnetism lie in the fact that electret phenomena can originate from electric monopoles as well as dipoles and they can occur to some small degree in any dielectric.

3.5.5 D.C. CONDUCTIVITY OF CRYSTALLINE SOLIDS

The low field d.c. conductivity of ceramics is usually dominated by grain boundaries. It is normally enhanced by conduction along them but occasionally is reduced by interfacial charge build up. This makes it difficult to reproduce measurements between similar specimens, even within magnitudes. The d.c. conductivity of crystalline polymer dielectrics is much more reproduceable because sharp grain boundaries are effectively absent. Values broadly correlate with permittivity: the higher the permittivity, the higher the d.c. conductivity, suggesting ionic mechanisms in most cases.

3.5.6 LOSS TANGENT OF CRYSTALLINE SOLIDS

The loss tangent of crystalline simple dielectrics and electrets can be due to ions or electrons. It is impurity controlled at most temperatures for all but the purest single crystals. Polycrystals are usually more lossy, if only by virtue of interfacial polarisation at crystallites and charge transport down grain boundaries.

The loss tangent of para- and ferroelectrics can be impurity controlled below 100°C or so, but at higher temperatures it tends to be dominated by the ionic polarisation mechanisms that give them their high permittivity.

3.5.7 BREAKDOWN OF CRYSTALLINE SOLIDS

Single crystal dielectrics often have low breakdown strengths because they have poor resistance to thermal gradients. Polycrystals can be electrically and mechanically much stronger; indeed fine-grained polycrystalline ceramics are most useful for high voltage insulation. Surface irregularities are, of course, inevitable with polycrystals. These must be shielded with glazes in practical applications for ease of cleaning and prevention of local breakdown.

Inhibition of breakdown in solids does not usually take the form of providing atoms to 'mop up' electrons as in gases and liquids: atoms do not normally exist as such in solids. High breakdown strength is achieved by restricting the processes that produce electrons by, for instance, choosing pure materials with a large forbidden bandwidth.

FURTHER READING

ALSTON, L. L., *High Voltage Technology*, Oxford University Press (1968)

BEYNON, J., *Conduction of Electricity in Gases*, Harrap, London (1972)

BOLTON, B., 'Discharges in Insulation', *Electl Times*, **155**, 61 (1969)

BRUIN, P. F. (Ed.) *Plastics for Electrical Insulation*, Interscience, London (1968)

BUDWORTH, D. W., *An Introduction to Ceramic Science*, Pergamon Press, Oxford (1970)

DAKIN, T. W. and BERG, D., 'Theory of Gas Breakdown', *Progress in Dielectrics*, **4**, 153, Iliffe, London (1962)

DENNIS, W. H., 'Properties and Uses of Electroceramics', *Electl Times*, 43, 14 May (1971)

FELICI, N. J., 'The De-Ionisation of Strongly Polar Liquids', *Brit. J. appl. Phys.*, **15**, 801 (1964)

GEVERS, M., 'The Relation between the Power Factor and the Temperature Coefficient of the Dielectric Constant of Solid Dielectrics', *Philips Res. Rep.*, **1**, 279 (1946)

HARROP, P. J., 'Temperature Coefficients of Capacitance of Solids', *Jnl Mater. Sci.*, **4**, 370 (1969)

HERBERT, J. M., 'Survey: Ferroelectric Ceramic Materials', *Component Technol.*, **4** (8), 17 (1971)

Index to the Literature on Ferroelectric Materials and Ferroelectricity, Plenum Press, London (1971)

ISARD, J. O., 'The Mixed Alkali Effect in Glass', *J. Non-Cryst. Solids*, **1**, 235 (1969)

KINGERY, W. D., *Introduction to Ceramics*, Wiley, New York (1960)

KOK, J. A., *Breakdown of Insulating Liquids*, Philips Tech. Library (1961)

KRASUCKI, Z., 'Breakdown of Liquid Dielectrics', *Proc. R. Soc.*, **294**, 393 (1966)

LAMB, D. R., *Electrical Conduction Mechanisms in Thin Insulating Films*, Methuen, London (1967)

LLEWELYN JONES, F., *Ionisation and Breakdown in Gases*, Methuen, London (1966)

LOEB, L. B., *Electrical Coronas*, University of California Press (1965)

MERZ, W. J., 'Ferroelectricity', *Progress in Dielectrics*, **4**, 101, Iliffe, London (1962)

OWEN, A. E., 'Electric Conduction and Relaxation in Glass', *Progress in Ceramic Science*, **3**, 77, Pergamon Press, Oxford (1963)

PAPOULAR, R., *Electrical Phenomena in Gases*, Iliffe, London (1965)

POPPER, P., 'Non-oxide Ceramic Dielectrics', *Progress in Dielectrics*, **1**, 219, Iliffe, London (1959)

SESSLER, G. M. and WEST, J. E., 'Studies of Electret Charges Produced on Polymer Films by Electron Bombardment', *Poly. Lett.*, **7**, 367 (1969)

SMYTH, C. P., *Dielectric Behaviour and Structure*, McGraw-Hill, New York (1955)

STEVELS, J. M., 'The Electrical Properties of Glass', *Handbuch der Physik*, **20**, 350 (1957)

SUTTON, P. M., 'Space Charge and Electrode Polarisation in Glass', *J. Am. Ceram. Soc.*, **47**, 188 and 219 (1964)

WAYE, E. E., *Introduction to Technical Ceramics*, Maclaren, London (1967)

ZAKY, A. A. and HAWLEY, R., *Dielectric Solids*, Routledge and Kegan Paul, London (1970)

4

Uses for dielectrics

4.1 INTRODUCTION

Various uses for dielectrics have already been mentioned in passing, but it would be useful at this stage to take a broader view of the applications of dielectrics, identifying the primary uses.

Dielectrics are most generally used as ordinary insulators; but their other unique properties are increasingly employed in various electronic devices. These devices can be divided into passive devices, being those which broadly obey Ohm's law, and their counterparts, active devices. The following examples are in no way comprehensive but merely indicate the principles behind some of the most important dielectric devices.

4.2 GENERAL INSULATION

The basic principle to note with general insulation is that the impedance is important as it is a measure of the leaking away (wastage) of the signal or power transmitted by the device. In view of this one must always remember that the impedance of a dielectric at d.c. is equal to its d.c. resistivity, whereas its impedance at a.c. is a function of that quite distinct variable, the permittivity. Further, when one uses layers of different dielectrics the d.c. impedance is simply the sum of the various resistances, whereas the a.c. impedance is in the simplest case proportional to the sum of the reciprocal permittivities. Accordingly with a.c. one quickly gets into the field of diminishing returns if one attempts to influence overall impedance by changing the permittivity of a single layer. This is not true of d.c. impedance.

Similarly, the heating the dielectric must tolerate is for d.c. a function of d.c. resistivity, but for a.c. it is a function of loss tangent. Bearing in mind that d.c. resistivity is not usually related to loss tangent, one realises again that quite different materials may be appropriate for a.c. as against d.c. use.

In practical systems, breakdown occurs earlier under a.c. conditions due to discharges in voids or created bubbles. Another point that contributes to this is that a.c. conductivities are always higher than d.c. conductivities at low fields and, for this reason alone, the d.c. rating of a given layer of insulation can usually be appreciably higher than the a.c. rating.

Of course, with any practical application of general insulation there are dozens of non-electrical considerations to take into account. Chemical deterioration can be important. Resistance to humidity, cost, availability of materials and so on have to be

Table 4.1 PROPERTIES OF SOME POLYMERS USED IN INSULATION

	Thinnest gauge available (μm)	*Lowest loss tangent in use*	*Permittivity*	*Upper working temp* (°C)
Polyimide	25	0·01	3·4	450
Polytetrafluoroethylene	12	0·00001	2·1	280
Polysulphone	3	0·01	3·1	160
Polyethylene terephthalate	2·5	0·01	3·0	140
Polycarbonate	1·5	0·003	3·0	125
Polypropylene	6	0·0001	2·3	100
Polystyrene	8	0·0001	2·6	75
Polyethylene	12	0·00005	2·2	70
Polyphenylene oxide	25	0·001	2·6	175
Paper + oil	2 × 5	0·01	4–6	85

considered. It is therefore no surprise that the material with something less than the best available electrical properties tends to be used in a given application.

Electrical insulation for temperatures above 500°C or frequencies above 1 GHz usually consists of ceramics because other materials are poor insulators under these conditions. Inorganic glasses and organic polymers are applied below this temperature. In particular the application of polymers is increasing steadily, as these can be tailored economically to meet ever more stringent specifications. Two aspects of this are shown in *Table 4.1*. It can be seen that the rather polar polymers, identified by their higher permittivity and dissipation factor, can be manipulated into rather thinner films and/or work at higher temperatures. Therein lies their attraction.

Gaseous insulation employs the freons (e.g. CF_2Cl_2, $CHClF_2$) fluorcarbons (e.g. C_3F_8, C_4F_{10}) and fluorides (SF_6, SeF_6) to 'mop up' electrons, or dry air or inert gases in the less exacting circumstances. The primary liquid insulants are organic, the best being fluorcarbons and silicones (again to mop up electrons) whereas waxes, mineral oils, etc., are used for less demanding applications.

4.2.1 CABLES

Ever since Le Sage constructed the first practical electric telegraph in 1774, the provision of adequate insulation has been one of the major problems of power and signal transmission.

In many cases, cables have to be pliable even at low temperatures. Power cables may have to have forced cooling by circulating air or oils. Other cables such as those used in aircraft may have to be as light-weight as possible. Some important electrical factors in the choice of dielectric for various categories of cable are given in *Table 4.2*.

Table 4.2 MAIN ELECTRICAL FACTORS INFLUENCING THE DESIGN OF CABLES

Type	*Properties of main concern*	*Present dielectric*
A.C. power cable e.g. 100 kV	Low loss, a.c. electric strength, i.e. heat and breakdown	Paper with thin transformer oil or dodecyl benzene or Paper with viscous mineral oil with N_2 overpressure
A.C. signal cable e.g. 1 V	Low permittivity, i.e. loss of signal	Solid polyethylene or Foamed polyethylene
D.C. power cable e.g. 100 kV	High and temperature invariant d.c. resistivity, i.e. loss of power, heating	Paper with transformer oil

The wall of a power cable is usually built up of several layers of paper or plastic 50–100 μm in thickness. These may be impregnated with oil before or after assembly to inhibit discharges in voids. The layering tends to ensure that any flaw in one layer is unlikely to line up with one in another and the integrity of the composite is far better than that of a single thick layer. Although this is basically the construction used in the last century, the purity of the paper has been improved until it is almost pure cellulose and its thickness and density is now very uniform. This has resulted in very great improvements in electrical properties. Gaseous overpressure may be created by direct pressurisation or use of an outer gas sheath. A simple solid sheath of polymer would be better than the above complexities but this calls for application of the polymer without creating any voids whatsoever, at reasonable cost. This has been attempted with little success for twenty years.

For all but the highest voltages, the cost of the cable now equals the cost of installing it. There is therefore less incentive for further

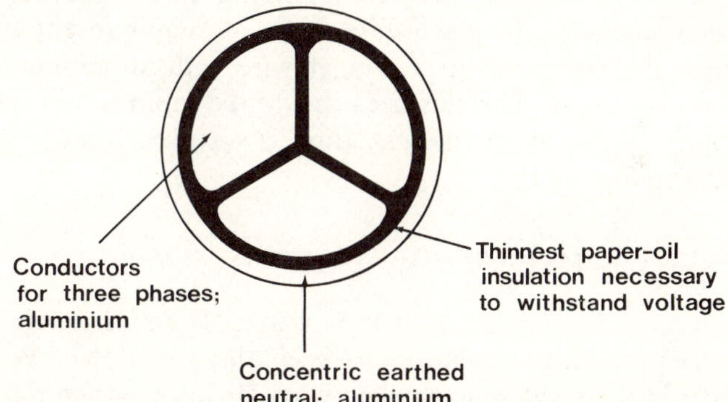

Figure 4.1. A modern medium power a.c. cable in cross-section. Use of a paper consisting of polypropylene fibres is thought by some to be a more realistic next stage of development

reduction in the cost of the dielectric, which is a small part of the whole anyway. Reliability is important, and oil impregnated paper is still difficult to beat in this respect. A modern medium power cable is illustrated in Figure 4.1 where the paper is the thinnest capable of withstanding the electrical and thermal gradients involved. For power transmission superconducting cables may one day become commercially viable and here the problem will not be one of heat dissipation but rather of temperature retention and low thermal conductivity will be of prime importance.

Insulation for cables used at temperatures above 100°C (e.g. in underfloor heating, nuclear reactor instrumentation) can be provided by various ceramics such as MgO (magnesia) powder or high purity polycrystalline Al_2O_3 (alumina) beads, separating a central wire from an outer metal tube. Although magnesia is very commonly used for pliable copper cables even at 20°C it must be noted that it is very sensitive to humidity should a leak occur at any terminations. Also spurious pulses and currents can be created when one is measuring high impedance phenomena as in the operation of radiation detectors.

4.2.2 TRANSFORMERS

With transformers and various forms of switchgear and generators the heat dissipation problem is particularly acute and a common way of solving it is to insulate with a transformer oil (the same mineral oil used in cables) which has as its main attribute a very low viscosity facilitating convection cooling. Its low freezing point also recommends it for use out-of-doors. A major recent advance has been the replacement of copper wire with aluminium foil in many transformers. This has been due to the rapid rise in the price of copper. In this event the insulation is switched from lacquer to plastic or paper film.

4.2.3 ELECTRIC MOTORS

Small electric motors, as used in power tools and domestic appliances, require thin insulation to keep the size down. Windings therefore still consist of enamelled wire. However, when the device is overloaded, the windings heat and the most common form of failure is electrical breakdown of the enamel insulation. There is much scope for improvement of the insulation in all electric motors.

4.3 PASSIVE DEVICES

4.3.1 CAPACITORS

The whole essence of a capacitor is the capacitance afforded by a thin layer of dielectric. The first practical capacitor was the 'Leyden

Table 4.3 TYPICAL REQUIREMENTS FOR NON-POLAR CAPACITORS WITH EXAMPLES OF APPROPRIATE DIELECTRICS

Use	*Capacitance*	*Voltage*	*Upper Temperature* (°C)	*Upper limit of loss tangent* (1 kHz, 20°C)	*Appropriate dielectrics*	*Particular benefits*
Interference suppression	0·5 μF	250 V a.c.	100	0·01	Paper + solid epoxy polymer impregnant	Cheap
Power station equipment	1000 pF	1 MV d.c. impulse	70	0·02	Paper + viscous transformer oil	Prevents impurity ions moving
	1000 pF	400 kV a.c.	70	0·02	Paper + thin transformer oil	Convection cools
Pulsed energy source, e.g. welding, laser pumping	100 μF	1 kV pulse	70	0·01	Paper + castor oil	Withstands discharges
Radio transmitters	1000 pF	10 kV radio frequency	100	0·01	Mica stacked in thin transformer oil	Non-inductive construction; prevents discharges
Power factor correction of industrial electric motors	10 μF	440 V a.c. surge		0·01	Paper or polypropylene + chlorinated diphenyl	Small size, high reliability
Tuning circuits	1000 pF	50 V d.c.	75	0·0005	Polystyrene or polypropylene film	Stable, accurate
Coupling in electronics	0·5 μF	50 V d.c.	125	0·01	Polycarbonate or polyethylene terephthalate film	Cheap, small
R.F. heating circuits	1000 pF	2 kV radio frequency	150	0·03	Ferroelectric and/or simple ceramics	Reliable, withstands heat
Fluorescent light ballast	8 μF	250 V a.c.	85	0·0005	Polypropylene film	Will not discharge liquid on failure

jar' invented by Van Musschenbroek in 1746. It had a glass jar as dielectric. Capacitors now take a wide variety of forms but they usually consist of electrode–dielectric–electrode sandwiches that are either stacked up or wound up. Since this is a symmetrical arrangement, such capacitors are known as '*Non-polar Capacitors*' a little confusingly since polar plastics are frequently used in their construction.

Most capacitors are used in electronics where they commonly have to withstand d.c. voltages up to 10 kV or so or a.c. voltages up to 250 V. Some are the size of a pinhead. Nevertheless, about a third of all capacitors used are applied in electrical engineering—for instance, starting motors and correcting the power factor of inductive loads such as fluorescent lights in order to save electricity charges. Electrical engineering calls for capacitors working at up to 2 MV a.c. or d.c. Types used in power stations may be as large as 10 m high.

In contrast to cables, the *lowest* a.c. impedance is important and one might expect high permittivity materials in use. However, other factors such as availability of thin layers, breakdown strength, price and reliability usually outweigh permittivity considerations. Oil impregnated paper, one of the most popular cable dielectrics, is also among the most popular capacitor dielectrics, although the wider variety of applications of capacitors calls for a wider selection of materials. *Table 4.3* illustrates how this is typically resolved.

Temperature requirements are sometimes extreme. Accordingly, in modern aircraft, capacitors to restart the engines in flight may have to operate at $+350°C$ whereas capacitors in some of the electronic circuits may drop to outside temperatures approaching $-55°C$.

The dielectric thickness is chosen so that the operating voltage is a half to a tenth of the breakdown voltage. The inevitable flaws and weak points in the dielectric preclude higher stresses. Thus mica, paper, ceramic and plastic usually support 10^6–10^8 V m^{-1} in practice and further radical reductions in capacitor size are unlikely. The one exception is low voltage electronics where even 2 μm polycarbonate is wastefully thick. In the future, progress will lie in new synthetic materials—polypropylene with or without oil replacing paper because of improved properties at a similar price, and so on.

The exception, low voltage electronics, will see deposited dielectric thin (0·2 μm) films replacing self-supporting films, and integrated circuits eliminating many capacitors in electronics anyway.

It is interesting that, even in energy storage and pulsed energy source applications, the ferroelectric has limited application because its crystalline structure results in low breakdown strength (Section 3.4, Breakdown). Consequently although energy stored is equal to $0{\cdot}5\,CV^2$ (Chapter 1, Energy Storage) working voltage is proportional to breakdown strength and this outweighs permittivity (capacitance) considerations. Low permittivity papers and plastics are therefore best for energy storage.

Fortunately the capacitor manufacturer has an ingenious technique available for producing far greater capacitance per unit volume where this is of primary concern and tan δ and $R_{d.c.}$ are less important. He operates insulating layers as thin as 0·02 μm simply by contacting them with conducting liquids—electrolytes—and applying a d.c. voltage during the operation in such as direction that

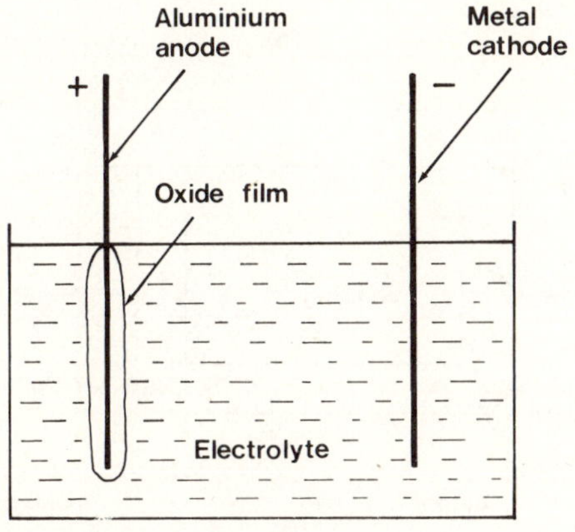

Figure 4.2. Schematic illustration of anodisation. The electrolyte could be aqueous glycol borate. The oxide film grows about 12 Å for every d.c. volt applied

flaws in the film are continually healed by corrosion of the supporting metal. Capacitors made in this way fall into a second category known as '*Electrolytic Capacitors*'. The creation and healing of the film by an electrically controlled chemical reaction is known as anodisation or forming. It is illustrated in Figure 4.2. A simple electrolytic capacitor is made by anodising an aluminium or tantalum foil and winding it up with a paper 'separator' and aluminium counter-electrode as in Figure 4.3, the whole being soaked in a suitable electrolyte. Usually, the surface area of the metal is increased by a controlled electrochemical pitting process

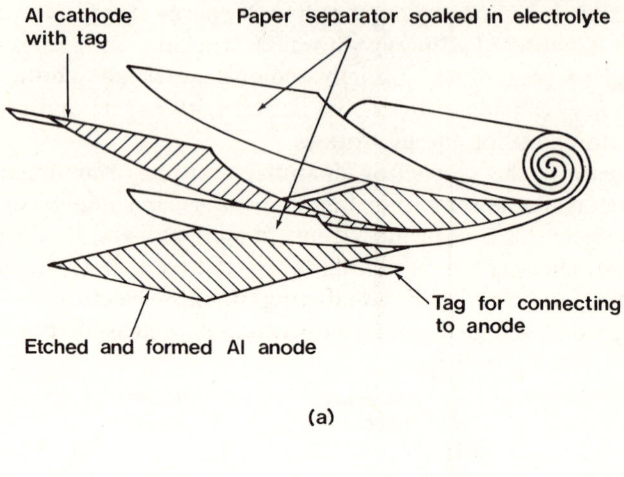

(a)

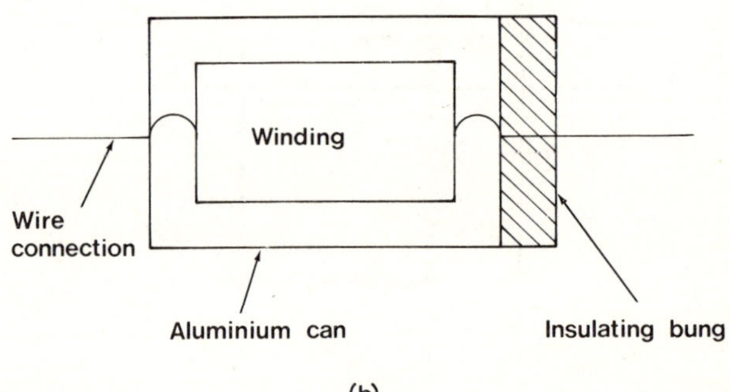

(b)

Figure 4.3. Typical electrolytic capacitor. The cathode may be connected to the metal case. The anode connection comes out of the case through an insulating bung

prior to anodisation. For a given capacitance, this can reduce the size of capacitor by a factor of ten. Some electrolytic capacitors have a solid electrolyte such as MnO_2 instead of a liquid one.

The dielectric in electrolytic capacitors is Al_2O_3 or Ta_2O_5 glass formed on Al or Ta respectively. As expected from Section 3.4, these films have high breakdown strength. In contrast to the non-polar construction, the dielectric can be operated near to breakdown, 4–5 × 10^8 V m^{-1}, but it does have rather inferior electrical properties.

Aluminium forms the less chemically stable oxide and is used in cheap and very high capacitance applications such as computer power supply smoothing (up to 1 F). It is by far the cheapest capacitor per microfarad of any available. However inflated loss tangent, due to series L and R effects, limits its use for expulsion of energy, radio frequency, etc.

Tantalum is expensive but gives highly reliable capacitors used exclusively in electronics up to 3000 μF. It can nevertheless be cheaper per microfarad and much smaller than any non-polar equivalent. Decoupling is a typical use.

Variable capacitors conveniently employ air, ceramic or, for high voltage, pressurised gas or vacuum (cf. Paschen plot, Section 3.2) since low capacitance and precision is usually called for.

To give a general view, some pertinent electrical properties of the main categories of capacitor are summarised in *Table 4.4*.

Table 4.4 TYPICAL ELECTRICAL PROPERTIES OF THE MAIN CLASSES OF CAPACITOR

Type	*Dielectric*	*Capacity* (μF)	tan δ (1 kHz)	$\tau = RC$ (s)	*Temp. coeff.* γ_c (ppm/°C)
Non-polar	solid or solid + liquid	10^{-6}–100	10^{-4}–10^{-2}	10^4–10^6	−150 to +300 (ceramics down to −20000)
Electrolytic	solid	0·1–10^6	10^{-2}–0·2	1–10^3	10^3–10^5
Variable	gas or solid	10^{-6}–10^{-3}	10^{-4}	10^6	zero

Finally, in the capacitor business the 'Equivalent Series Resistance' of a capacitor is frequently quoted. This is related to the true (parallel) resistance and capacitance by the equations in Appendix 2.

4.3.2 RESISTORS AND STRAIN GAUGES

Although capacitors are the major form of passive device relying on dielectrics as an integral part of their structure, there are various fascinating applications of dielectrics in resistors. For instance if one wants a thin film miniature resistor of several megohms value one can make this by vacuum depositing a film of a few hundred nanometres thick of cermet, that is a dielectric with islands of metal interspersed in it as in Figure 4.4(a). The islands may be 2 nm across. The conduction along this layer consists of electrons passing

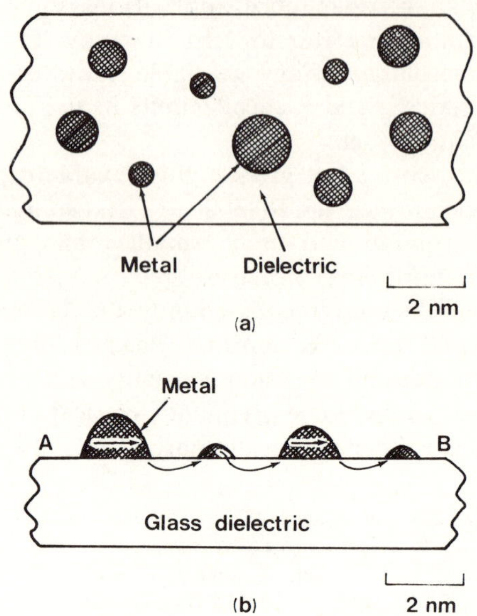

Figure 4.4. (a) A cermet and (b) an island structure thin film showing the electrical conduction path from A to B. The limiting steps are conduction through the dielectric

from island to island, the limiting step being conduction in the dielectric.

A two-dimensional equivalent of the cermet can be made by evaporating a tiny amount of metal onto a glass slide. This condenses in the same configuration as drops of water on a window pane. Conduction across this surface is then from island to island as in Figure 4.4(b). This arrangement is also used for megohm resistors. However, the conduction, being by electron tunnelling over such small distances (see Figure 3.11 and associated discussion), is extremely sensitive to island separation and any bowing of the substrate causes a large change in current. One therefore has an excellent strain gauge as well.

4.4 ACTIVE DEVICES

In this book we shall ignore those active devices based on ceramics and glasses that exhibit switching phenomena, rectification, photoconductivity and so on because their behaviour is on a par with semiconductors, indeed they typically exhibit the active properties required only when they are 'doped' with impurity and have a forbidden bandwidth of less than 3–4 eV.

4.4.1 FIELD EFFECT TRANSISTORS

More important in a book on dielectrics is the field effect transistor. This is a cheap device useful, for instance, in television sets where its slow speed is no disadvantage. Examples are illustrated in Figure 4.5. Here the very heart of the device is an insulating 'gate' usually formed by careful corrosion in 1300°C steam of high purity silicon

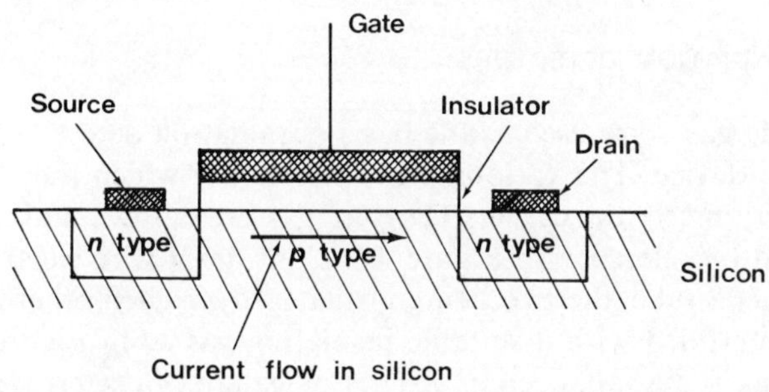

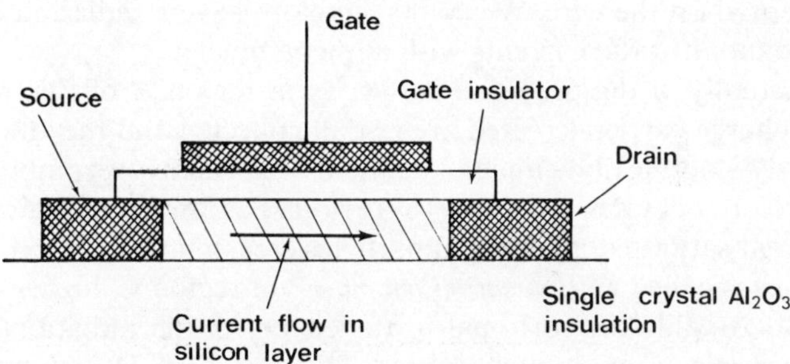

Figure 4.5. Two types of field effect transistor. These may be only a few μm across. They are commonly built into arrays with many hundreds of other similar components, the complete 'integrated circuits' being approximately 2 × 2 × 0.5mm

to give a silica (SiO_2) glass film. This then has a thin layer of aluminium evaporated onto it. The device operates by the field applied at the gate modifying the transverse flow of electrons in the semiconductor. Of vital interest here are the interfacial states—traps and generators of electrons caused by ionic impurity and imperfections in the insulator at the insulator–semiconductor interface. Of course it is also important that the gate be reasonably thin—say 0·5 μm—yet have a high breakdown voltage and low d.c. conductivity. Trouble in the insulator is generally caused by water or sodium ion contamination. The latter can be minimised by incorporating a layer of silicon nitride on top of the insulating layer. Glassy layers of silica and silicon nitride a few hundred nanometres thick are used for general insulation in integrated circuits when conducting layers pass over each other and on top of the finished device. They exhibit extremely high breakdown strength, low loss and high d.c. resistance as expected from the principles laid down earlier in Sections 3.1 and 3.4.

4.4.2 RADIATION DETECTORS

Dielectric gases are used as the basis of radiation detectors, a form of active device. The various principles under which they operate cover very nicely the earlier theory of gas breakdown and are now described by reference to Figure 4.6. Consider a wire electrode led into a metal tube, the two being insulated from each other and the tube being filled with a suitable insulating gas as in Figure 4.6(a). If ionising radiation, such as X-rays or gamma rays, creates a few free electrons and ions, either of these can be collected by a suitable potential on the wire. We could therefore have a radiation detector if we monitor the currents with a 'picoammeter'.

Actually, if the potential is low, as in region A of Figure 4.6(b), the charge carriers created are not all collected and they move only slowly—ohmically—under the field and many recombine. One therefore operates the device in regions B, C or D. In region B, the current saturates, all charged particles created are collected, and the device is called an '*ionisation chamber*'. In region C, higher currents occur roughly proportional to the energy of the radiation and the device is a '*proportional counter*', and in region D, the gas breaks down by avalanching ionisation, pulses are monitored (a suitable gaseous additive quenches each pulse after an interval), and the

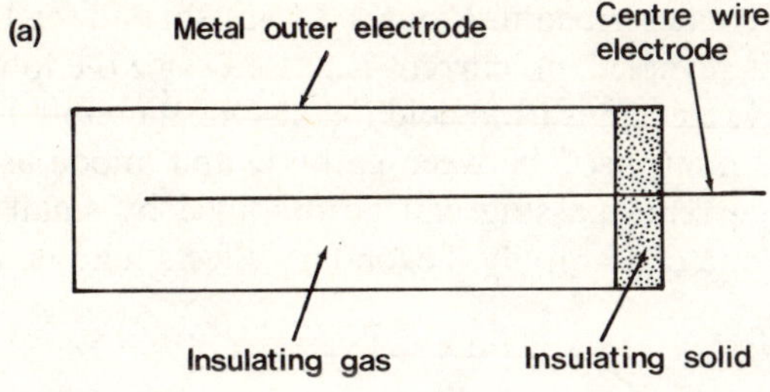

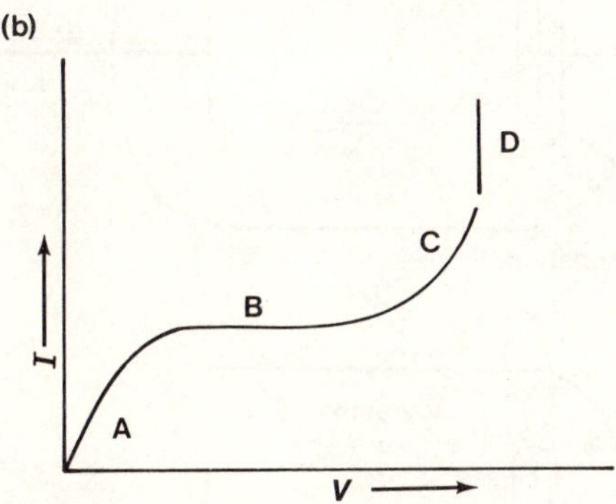

Figure 4.6. (a) A typical configuration for a radiation detector. (b) I/V curve showing regions of operation

device is a '*Geiger counter*'. In regions B and C one may monitor pulses of electrons or ions, the latter being slower moving, or one may measure integrated currents.

4.4.3 THERMIONIC VALVES

Another example of the use of dielectric gases as the basis of active devices is the thermionic valve which preceded the transistor and still has many uses. In the simplest case, a diode is made by having two metal electrodes in a vacuum as in Figure 4.7(a), making one of them negative and heating it. Electrons are emitted from the cathode, particularly if it is coated with a material of suitably low work function such as a heavy oxide [according to the Richardson–Dushman equation (3.13)]. Electrons are collected at the second electrode if it is made positive. The three modes of emission found in practice were described in Figures 3.16(a), (b) and (c). However, if the polarity is reversed, no current is passed since the low field used cannot remove electrons from cold metal. One therefore has a diode.

If a grid is interposed between cathode and anode as in Figure 4.7(b) large currents passing can be modified by small potentials applied and one can amplify. Secondary effects may be reduced in

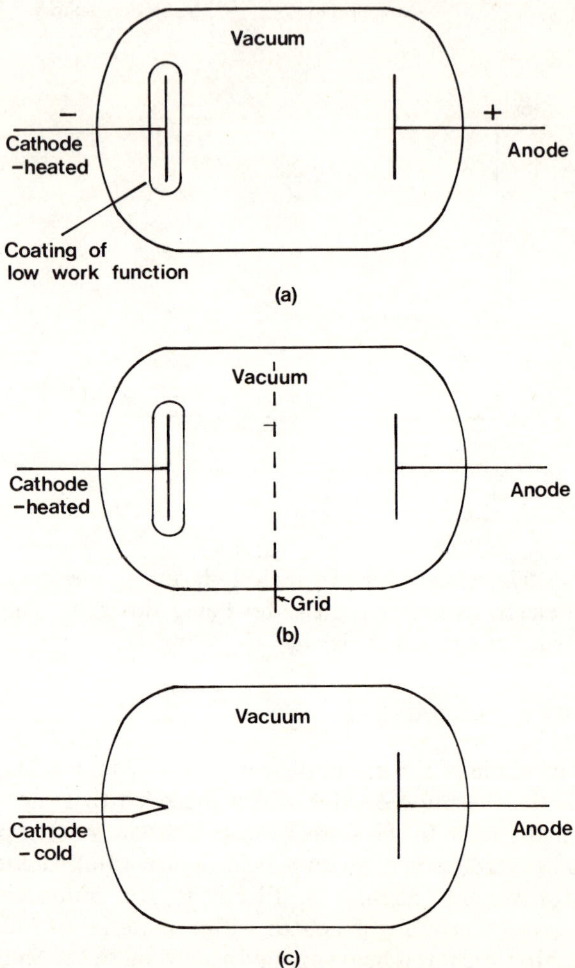

Figure 4.7. Various devices exploiting enhanced electron emission from a suitable electrode: (a) a thermionic diode; (b) a thermionic triode; (c) a cold diode

some cases by using an inert gas rather than a vacuum and by using more than one grid.

In some applications several kilovolts applied to a pointed cathode can produce a sufficiently high local field for Fowler Nordheim tunnelling [Figure 3.16(d)] even with a metal of relatively high work function. Heating is then unnecessary and a 'cold diode' as ir Figure 4.7(c) can be constructed.

4.4.4 ELECTRET DEVICES

From the practical point of view an electret can be used to produce a stable d.c. voltage bias of up to 100 V or more. It cannot be used as a current source, unlike a battery or a capacitor. Like a permanent magnet, it should preferably be shorted out during storage.

Sessler and West at Bell Telephone Laboratories and Sony, Japan, have used electrets in microphones. A metallised plastic foil electret

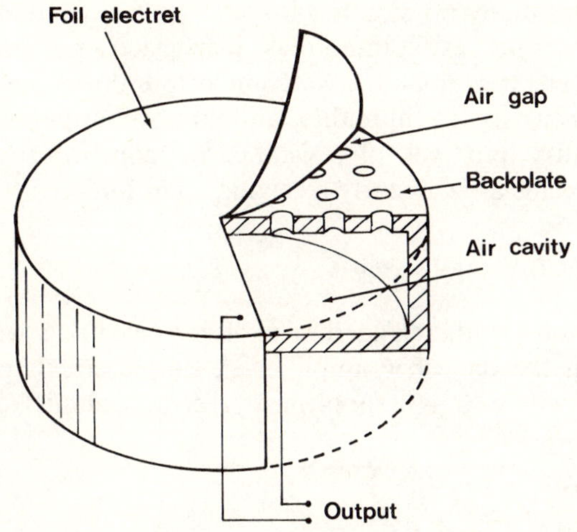

Figure 4.8. Illustration of an electret microphone. [*By permission of Bell Telephone Laboratories Inc.*]

is charged by a 10^6–10^7 V m^{-1} d.c. field at 200°C so that a positive charge is applied to one side while electrons are injected on the other. The foil is then built onto a backplate as shown in Figure 4.8.

As sound waves vibrate the foil, the electrostatic field across the air gap varies, producing a varying voltage at the output. Conventional techniques require a d.c. supply to monitor capacitance variation. The electret microphone is cheap, rugged and stable with temperature and time particularly because the construction compensates for any small charge decay over the years. Although such 'condenser' microphones only have one hundredth of the sensitivity of the carbon microphones now used in ordinary telephones, the electronic transmission techniques now being established may permit their higher fidelity to be used to advantage even here. This outlet for electrets could therefore become very large indeed.

Potential uses of electrets include electronic memories, electrophotography, dust collectors, gramophone pick-ups, radiation dosemeters, measurement of electrical potentials and prevention of blood clotting on artificial organs. Electrets may also be used in high impedance a.c. generators, gas ignition, analysis of mechanical vibration, earphones and for the deflection or acceleration of charged ions or electrons.

With these many possibilities it seems possible that electrets will one day become as common as piezoelectrics. Certain shortcomings of electrets must be overcome before this is the case. These include sensitivity to humidity and atmospheric pressure, irreproduceability and lack of resistance to temperature excursions. These problems are currently receiving attention.

4.4.5 DIELECTRIC AMPLIFIER

The variation of the capacitance of ferroelectrics with field is exploited in the dielectric amplifier. A simple circuit is shown in Figure 4.9 where C is a non-linear (i.e. non-ohmic) capacitor—

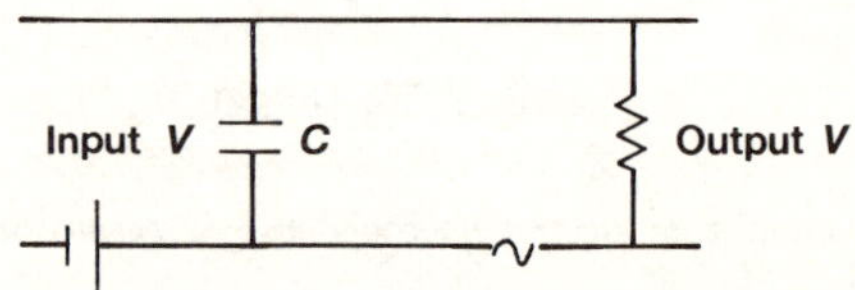

Figure 4.9. Simplified circuit of a dielectric amplifier

usually a ferroelectric behaving as in Figure 3.24. The capacitance is varied by the input signal and this produces a variation of current in the load and hence an amplified output voltage. The principle can also be employed in modulators and control devices.

4.4.6 PIEZOELECTRIC AND ELECTRO-OPTIC DEVICES

Piezoelectrics are usually used in the form of single crystals cut to a pre-determined shape with a particular crystal orientation. Quartz is particularly useful. Piezoelectrics are used as transducers, i.e. converting mechanical energy into electrical energy or vice versa. The principle is illustrated in Figure 4.10. Transducers are used to

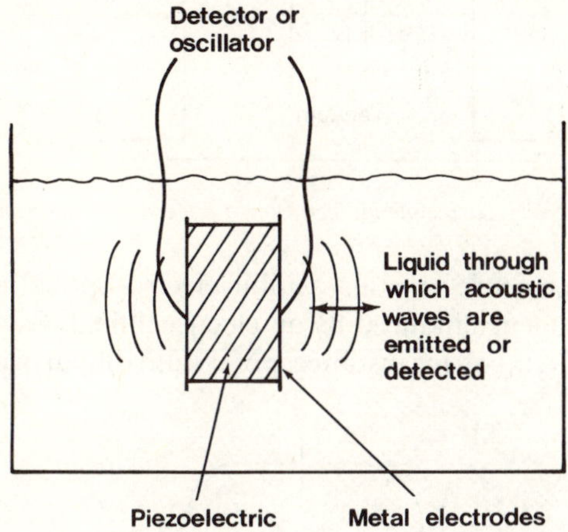

Figure 4.10. Principle of operation of a transducer. Gaseous or solid environments can also be used

generate ultrasonic waves for machining brittle materials, solvent cleaning, powder mixing, underwater detection and non-destructive testing (e.g. cracks in steel, position of the foetus in pregnant women) and in detection of the modulated waves in the last two applications. Sophisticated development of these properties has led to quite new principles for the construction of electrical components such as high voltage transformers and telephone filters.

If a piezoelectric crystal is energised in a vacuum as in Figure 4.11, i.e. not acoustically coupled, it can be made to resonate at a precisely calculable frequency. Piezoelectric resonators are used as frequency standards in aircraft communication systems. A piezoelectric crystal attached to a tuning fork acts as a low frequency filter. These have been widely used in telegraphy but are tending to be replaced by electronic circuits.

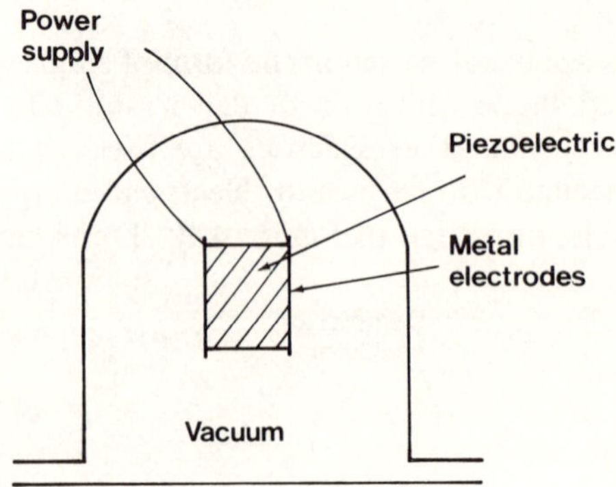

Figure 4.11. Principle of operation of a piezoelectric resonator

Many piezoelectric crystals exhibit electro-optical effects, particularly birefringence induced by an electrical field. Given a suitably transparent crystal of, for instance, potassium dihydrogen phosphate

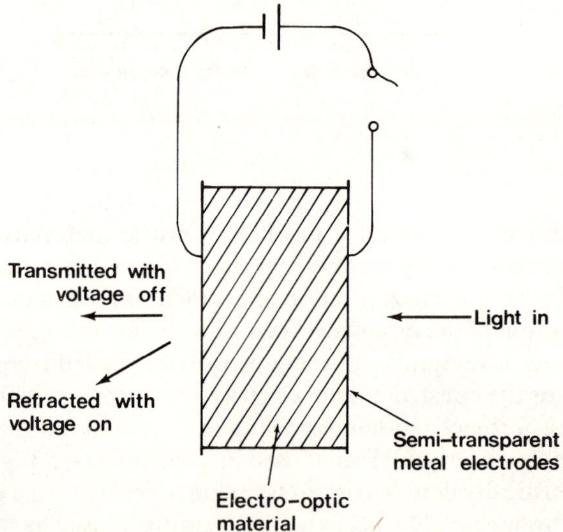

Figure 4.12. Principle of operation of an electro-optic device

(mentioned earlier as a ferroelectric–paraelectric) cleaved in an appropriate crystal plane and mounted as in Figure 4.12, laser or other beams of light can be switched at high speed from one direction to another. Such techniques should find uses in character recognition and computer memories.

Certain transparent liquids, known as 'liquid crystals', for instance methoxy butyl benzylidene aniline, and some ferroelectric solids can be made opaque or coloured when an electrical field is applied. This finds possible application in see-through map displays for pilots, shutters in photography, optical memory devices and window blinds. It may lead to colour television screens that hang on the wall like pictures. An easier application, that is already with us, is digital display on calculators. These modules operate on battery voltages and power consumption can be as low as 10^{-2} mW mm^{-2}. They are also used in thermography, the taking of temperature contours in medicine.

4.5 FUTURE TRENDS

Active and passive dielectric devices and even general insulation tend to be made to occupy less volume as time goes on. Firstly, this is because reliability is often increased, particularly due to a reduction in the number of interconnections. Secondly, the customer often wants to get more in a given space; one can think of hearing aids, portable video tape recorders, picture frame television sets, and almost anything that has to go into an aircraft cockpit. Thirdly, in some computer applications, the speed of response can be limited by the distance the electrons have to move. Finally, and often most important, where material costs are a large proportion of the whole, smaller size means lower price.

In electronics at least, discrete passive components are slowly being replaced by single devices that take the place of complete networks. Accordingly, resistor factories are tending to make complete digital–analogue convertors consisting of networks of resistors evaporated in one operation on a single glass slide using a suitable mask. Capacitor manufacturers are beginning to tap and wind single capacitors so that they simulate networks of capacitors, chokes and resistors such as pulse forming circuits in radar and filter networks. Hence, the customer who previously made an electrical filter out of capacitors, resistors and inductors can now ask the component manufacturer to achieve the function with a single device specially tailored from suitable dielectrics and conductors. In this way he can reduce component cost, cost of mounting and size, and increase reliability.

Newly discovered properties of existing dielectrics are also being exploited. What was essentially a polycarbonate capacitor has been made to exploit the electret properties of the material and used in a microphone as already described. With the field effect transistor, the use of thin gates is giving switching effects that may be the basis of improved computer memories. Progress will depend on a more thorough understanding of the dielectrics employed.

In future years, oil-impregnated or gas-pressurised plastic will replace oil-impregnated paper in cables and capacitors due to superior quality at a lower price. Polyethylene, polypropylene and polyphenylene oxide are of interest for power cables, polypropylene for power capacitors. However, ceramics and inorganic glasses will retain most uses. Their development has scarcely begun. Gaseous insulation will increasingly be replaced by solid materials, in devices like radiation detectors, to improve size and reliability.

As the non-electrical properties of synthetic materials are brought under control they will increasingly be used to replace natural dielectrics. These synthetic dielectrics will consist almost entirely of elements from the first three periods of the Periodic Table for the basic reasons given earlier in this book.

FURTHER READING

ALSTON, L. L. (Ed.) *High Voltage Technology*, Oxford University Press (1968)

ANDERSON, J. C., *Dielectrics*, Chapman and Hall, London (1964) A

BARNES, C. C., *Power Cables*, Chapman and Hall, London (1966)

BLATCH, H. R., 'Piezoelectrics: Cornerstone of Transducer Design', *Design Engng*, 55, April (1972)

BOGORODITSKII, N. P. and PASYNKOV, V. V., *Radio and Electronic Materials*, Iliffe, London (1968)

CAMPBELL, D. S., 'Survey: Electrolytic Capacitors', *Component Technol.*, **4** (1971)

CAMPBELL, D. S. and MORLEY, A. R., 'Electrical Conduction in Thin Metallic, Dielectric and Metallic–Dielectric Films', *Rep. Progr. Phys.*, **34** (4) (1971)

CHOPRA, K. L., *Thin Film Phenomena*, McGraw-Hill, New York (1969)

COBBOLD, R. S. C., *Theory and Application of Field Effect Transistors*, Wiley, New York (1970).

DUMMER, G. W. A., *Connectors, Relays and Switches*, Pitman, London (1966)

GLANG, R. and MAISSEL, L. I. (Eds.) *Handbook of Film Technology*, McGraw-Hill, New York (1970)

GORE, W. (Ed.) *Microcircuits and Their Applications*, Iliffe, London (1968)

GROSS, B., 'Electrets'. *Endeavour*, **30**, 115, Sept. (1971)
HOLLINGSWORTH, D. T., 'Cable Development', *Electl Times,* **155**, 49 (1969)
JACKSON. W. (Ed.). *The Insulation of Electrical Equipment.* Chapman and Hall. London (1954)
MAYOFIS. J. M.. *Plastics Insulating Materials.* Iliffe. London (1966)
PLOSS. R. S.. 'A Review of Electro-optics Materials. Methods and Uses'. *Opt. Spectra,* **3**. 63 (1969)
VANNER, K. C., 'Liquid Crystals and their Applications', *ERA J.* 13 (Autumn 1971)

5

Measurements on dielectrics

5.1 INTRODUCTION

It has already been observed that dielectric gases at appropriate pressures and with appropriate electrode separations can so nearly approach a vacuum in their properties that one is not greatly concerned with measurement of the infinitesimal d.c. and a.c. currents that may pass at low fields. Accordingly measurement of these systems has largely concentrated on breakdown. The greater part of this chapter will therefore be concerned with measurement of the electrical properties of solid and liquid dielectrics.

5.2 D.C. MEASUREMENTS

5.2.1 CONDUCTIVITY

Ideally conductivity measurements should always be made by four-terminal techniques where the current is passed between two terminals and two separate terminals act as voltage probes. This tends to exclude the spurious effects of contact potential and surface conductivity. Two useful arrangements are shown in Figure 5.1. With the arrangement as in Figure 5.1(a), calculation of conductivity is obvious. Figure 5.1(b) is appropriate to layers of constant

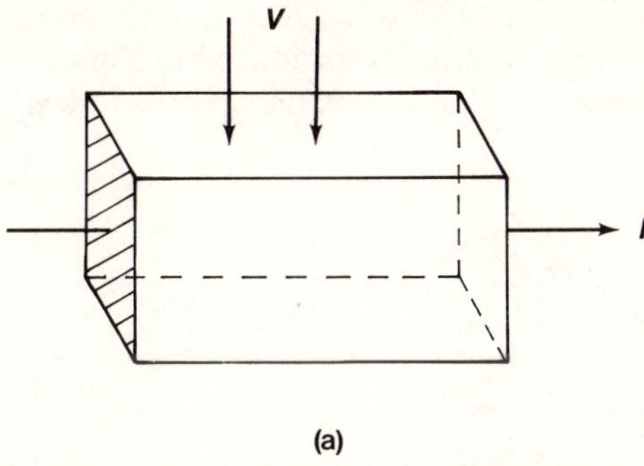

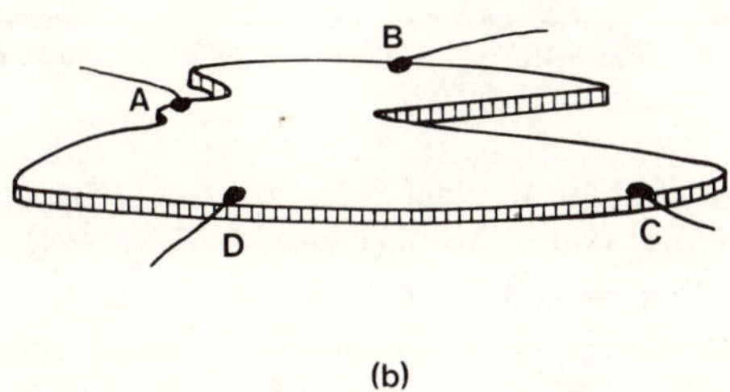

Figure 5.1. Illustration of (a) a typical four-terminal measurement, and (b) Van der Pauw specimen

thickness where the electrodes are effectively infinitely small. The analysis, due to Van der Pauw, is somewhat involved.

However, from the point of view of the experimentalist it can be reduced to very simple operations. We define resistance $R_{AB,CD}$ as the potential difference $V_D - V_C$ between the contacts D and C per unit current through the contacts A and B. We define $R_{BC,DA}$, etc. by analogy.

Van der Pauw showed that the bulk conductivity of the sample, σ, is given by

$$\frac{1}{\sigma} = \frac{\pi d}{\ln 2} \frac{R_{AB,CD} + R_{BC,DA}}{2} F \tag{5.1}$$

where F is a function of $R_{AB, CD}/R_{BC, DA}$ only and is conveniently taken from his general graph reproduced in Figure 5.2.

Figure 5.3 shows a Van der Pauw specimen designed in a more

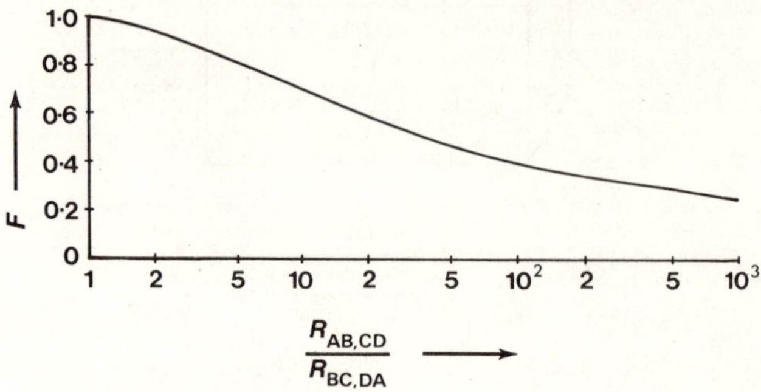

Figure 5.2. The function f used for determining the specific resistivity of the sample, plotted as a function of $R_{AB, CD}/R_{BC, DA}$. [After L. J. Van der Pauw, Philips Res. Rep., **13**, *1 (1958)*]

ideal fashion so that finite electrodes appear to be infinitely small. The Van der Pauw technique has been widely used on everything from uranium dioxide at 1800°C to sheets of plastic, and has made possible accurate measurements on odd shaped specimens where only unreliable two-terminal measurements had previously been possible.

When high fields are to be applied to anything other than thin films, potentials of several kilovolts have to be used and this requires that the specimen be designed to avoid gaseous or surface break-

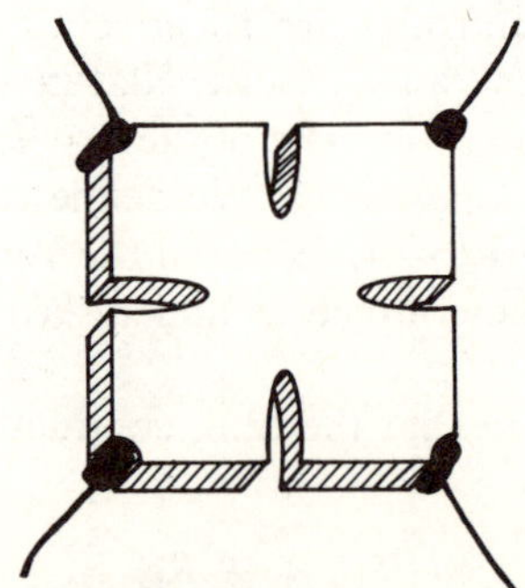

Figure 5.3. Van der Pauw specimen minimising the error due to finite electrode size

down around the sides. In this case a useful arrangement can be that shown in Figure 5.4 where external breakdown is avoided by making the appropriate paths rather longer than usual. This is a three-terminal method and may in principle be subject to stray

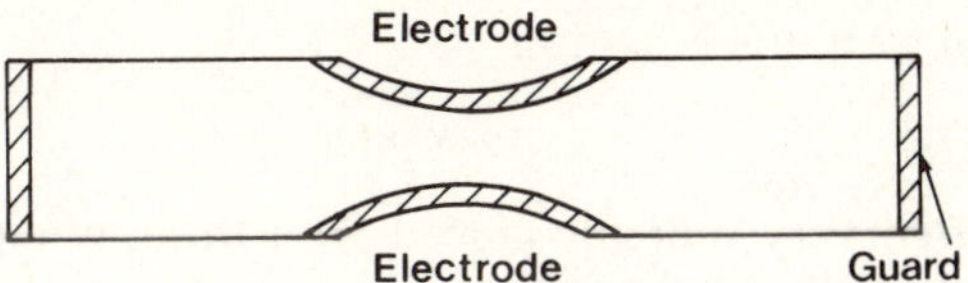

Figure 5.4. Electrode system commonly used for samples requiring high voltages

contact potentials at the electrodes. However, since these are usually only of the order of volts they are negligible. Surface leakage currents are leaked away by the guard and are not measured if neither measuring electrode is earthed.

As noted earlier in the book a dielectric will in practice give a d.c. conductivity that is a function of time, the current for a given applied voltage decaying with time. The decay usually represents a variety of processes (indeed all the processes shown at different a.c. frequencies) at different times, and it now becomes a problem to define what we mean by d.c. conductivity. Arguably we should take the conductivity at infinite time since by then all dipoles, representing a.c. phenomena, have oriented. In practical terms this is often done by taking the value when it is varying by only a few per cent per hour, or, more dubiously, taking the value after one minute. Since the resistivities of practical dielectrics may approach $10^{18}\ \Omega$ m one requires, even with thin-film samples, to measure 10^{12}–$10^{15}\ \Omega$.

Suitable instruments are the electrometer and the high input impedance valve voltmeter. The former is the most sensitive, and commercial versions, employing vibrating reed electrometer valves, will measure up to $10^{15}\ \Omega$ with only 100 V applied. Valve voltmeters must have an input impedance one hundred times the impedance to be measured. This usually makes them unsuitable for measurements above $10^{12}\ \Omega$. One must always remember that the d.c. conductivity of most materials increases above 10^6–10^7 V m^{-1} and values measured above this field, being non-ohmic, must have the field specified when they are collated. Of course, measurement of high field currents is much easier and d.c. power packs with voltmeters and microammeters often suffice.

5.2.2 MOBILITY

As with conductivity, we primarily concern ourselves here with measurement of the macroscopic low field (ohmic) mobility, known as the drift mobility. Basically this involves separating out the mobility μ from the equation

$$\sigma = ne\mu \tag{5.2}$$

This is often very difficult indeed with the tiny currents involved but a few techniques stand out.

Firstly one can determine *time of flight* by sudden injection of a

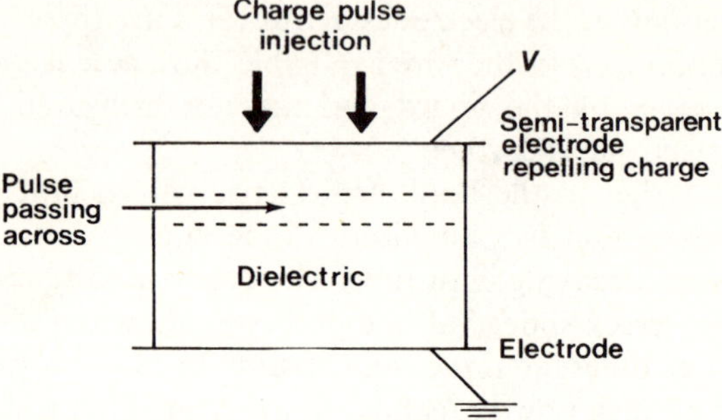

Figure 5.5. Conventional time of flight measurement of drift mobility. An electron beam is a typical source of charge

known small charge at one electrode on the dielectric as in Figure 5.5 and measurements of the transit time t under field E. For a specimen of thickness d, μ is given by

$$\mu = \frac{d}{Et} \tag{5.3}$$

The beauty of this method is that light, electron or ion beams can be used to inject a known charge carrier. Complications can arise when the charge transport is very slow and localised and space charge effects must be guarded against. Mobilities down to 10^{-8} m^2 V^{-1} s^{-1} have been studied in this way.

A useful development of the time of flight technique lies in measurement of the decay of surface charge on a dielectric by following the sequence in Figure 5.6. Here the surface charge (of

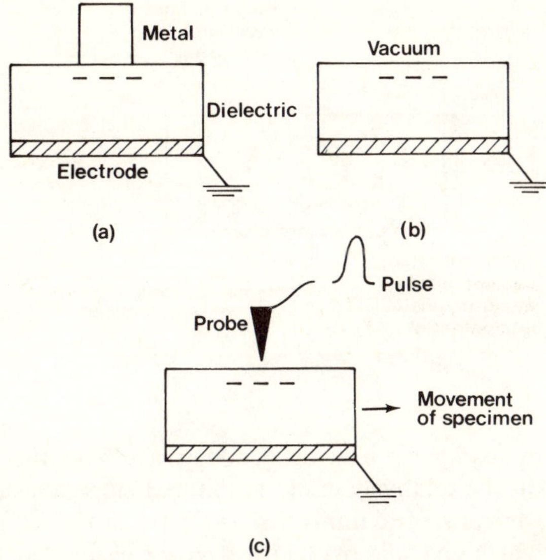

Figure 5.6. Davies method for determining mobilities of electrons in insulators. (a) An area of surface is charged by contact with metal of known work function; (b) the metal is removed; (c) the charge remaining after given time intervals is monitored by passing the insulator under a probe connected to an electrometer

electrons) could be created by metal contact as shown (this is what causes the band bending at interfaces, Section 2.2) or by irradiation with electrons or ultra-violet light. The latter two methods create electron-hole pairs by producing energy greater than the forbidden bandwidth. What one can then monitor, under suitable conditions, is the decay of this surface charge, not as a consequence of its drifting away across the surface, but either (*a*) at high temperatures by any pre-existing free carriers in the insulator moving up to the surface and neutralising it, or (*b*) at low temperatures, its drift into the bulk. One can check whether the injected charge is drifting away by surface conduction by seeing whether the pulse created on monitoring (Figure 5.6) is broadened. This technique is currently unexcelled for measuring the lowest mobilities, e.g. 10^{-14} m^2 V^{-1} s^{-1} in some solids at 20°C. It does not necessarily distinguish between ions, electrons and holes.

Thermoelectric effects can give an indication of mobility. The thermopower (Seebeck coefficient) is a measure of the number of free carriers independently of their mobility. Thus mobility can be

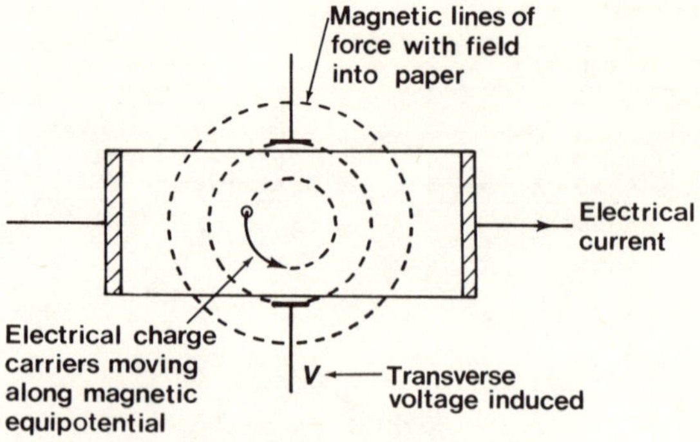

Figure 5.7. Illustration of the Hall effect

deduced by comparison with conductivity. The thermoelectric potential can be relatively easily monitored since non-metals give thermopowers of several millivolts per degree in Celsius in contrast to metals which give microvolts per degree Celsius. The sign of the charge at the cold junction gives the sign of the charge carrier.

A third method of measuring mobility is the *Hall effect* where a magnetic field is applied perpendicular to an electric current passing through a specimen. Figure 5.7 shows the monitoring of the transverse potential induced by the charge curving around the magnetic field equipotential. The sign of this transverse potential, as with thermopower, gives the sign of the charge carrier. Similarly again, the magnitudes of the quantities involved give a value for the number of conducting species and from this, and a value for the conventional d.c. conductivity, one can obtain mobility.

The great disadvantage of the Hall effect for our present purposes is that it is inappropriate for mobilities much less than 10^{-6} m^2 V^{-1} s^{-1} and therefore will not monitor ions or electrons moving at low fields in most liquid and solid dielectrics. It is worth noting that the Van der Pauw arrangement can be used for examining the Hall effect on irregularly shaped sheets of material.

The *Kerr effect* can be used to advantage particularly in measuring mobilities in liquids. This is the electro-optic effect by which most materials become doubly refracting under a sufficiently high electric field. The magnitude of the refraction is proportional to the electric field at that point. Thus if, for instance, a laser beam a fraction of a millimetre across is scanned across the liquid cell, the field distribution can be monitored between two electrodes. A typical arrangement can be that shown in Figure 5.8. To get a value of mobility one can create ions by having a radioactive electrode or by local irradiation with ultra-violet light or electrons. One can then follow

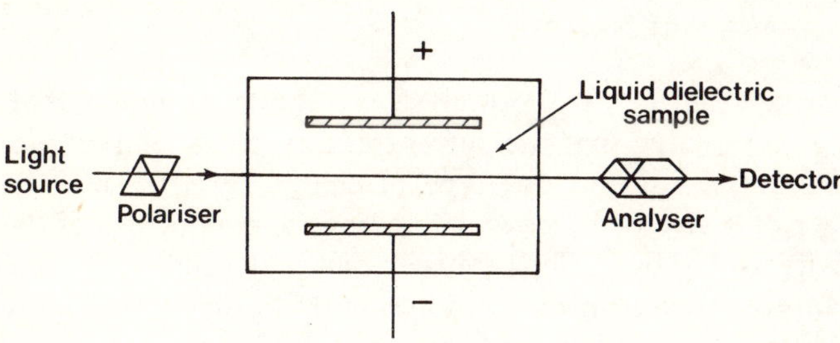

Figure 5.8. Illustration of the use of the Kerr effect for studying field distribution in a liquid

the progress of the charged particles created using the Kerr effect. Mobility can be calculated from rate of diffusion.

The mobility of ions in solids and liquids can be determined by *radio-tracer techniques* in many cases. A simple example is that of taking a block of material and coating a radioactive metal on one side. The block is then heated at a known temperature for a short time, t, so that the metal ions diffuse in somewhat. One then sections the block, determines the distance, d, that most have travelled and deduces the mobility from the relationship

$$d = \sqrt{(Dt)} \tag{5.4}$$

Table 5.1 COMPARISON OF TECHNIQUES FOR MEASURING MOBILITY

Method	*Advantages*	*Disadvantages*
Time of flight	Identification of charge carrier	Not suitable for lowest values
Decay of surface charge	Very low values are measurable	Poor identification of charge carrier in some cases
Thermoelectric effects	Simple; ions or electrons	Interpretation disputable
Hall effect	Ions or electrons	Insensitive; solids only
Kerr effect	Suitable for liquids	Complicated procedure
Radio-tracers	Simple; precise identification of charge carrier	Ions only; temperatures above 100°C

where the diffusion coefficient D is related to the d.c. conductivity and mobility by the Nernst–Einstein relation [equation (2.9)].

The various techniques for measuring mobility are compared in *Table 5.1*.

5.2.3 TRANSPORT NUMBER

To be quite rigorous in the analysis of conduction in materials, one must acknowledge that the current at any given temperature may not be entirely due to one type of charge carrier. The transport number of a given species is defined as the proportion of the conductivity caused by it. Thus if 50 per cent of the current is ionic, ions are said to have a transport number of 0·5 at that temperature. It is usually implied that low fields are used.

The standard technique is to heat the liquid or solid to a sufficiently high temperature for a substantial known low field current to be passed. Any electrolysis must be due to ions passing and this is measured by the weight gain of electrodes or possibly gas evolution. Faraday's Laws of Electrolysis give the electrolysis that should have taken place if the ionic transport number were unity and hence the true transport number can be calculated.

Limitations of this method are:

(1) that the electrodes may be partially blocking to ions;
(2) different mechanisms of conduction may occur at the elevated temperatures used.

5.3 A.C. MEASUREMENTS

There are a very wide variety of techniques quoted in the literature for measuring the a.c. electrical properties of materials. Three-terminal techniques are ideal, particularly with disc electrodes guarded as in Figure 5.9. This arrangement not only eliminates surface conduction but renders the electrical lines of force relatively parallel between the measuring electrodes. It can be shown that edge effects have a negligible effect on the capacitance if the guard–centre disc separation is small compared with the sample thickness.

The frequency ranges of the most popular circuitry are summarised in Figure 5.10 and their main advantages and disadvantages

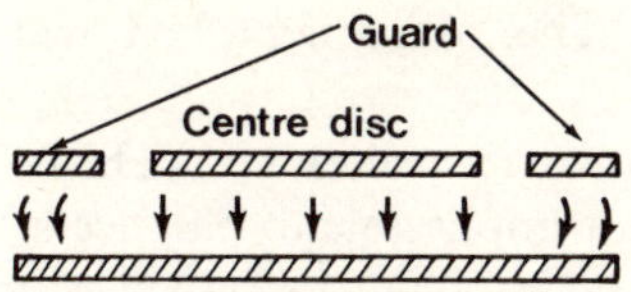

Figure 5.9. Side section of disc shaped electrode system showing effect of a guard ring

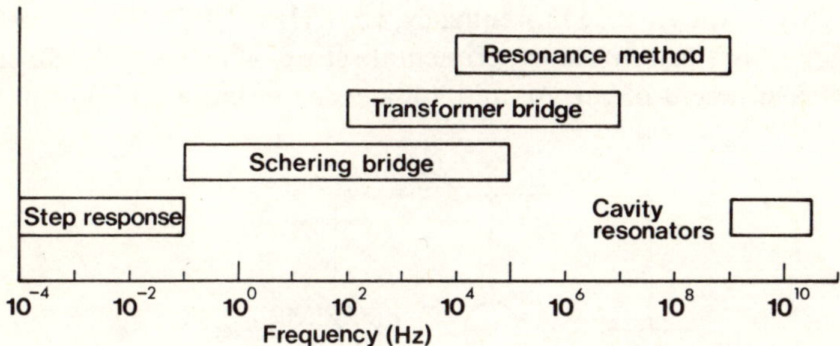

Figure 5.10. Frequency ranges of popular a.c. measurement techniques

Table 5.2 SALIENT FEATURES OF A.C. MEASURING CIRCUITS

Type	*Advantages*	*Disadvantages*
Step response	Simple apparatus; measures low loss	Not direct reading; ε' very inaccurate; takes some time
Schering bridge	Can be used from low to very high voltages and currents; direct reading	Strays can give errors; does not usually measure as low loss as transformer bridge
Transformer ratio arm	Strays are generally absent, three terminal measurements are simple; leads can be compensated for; direct reading; very low loss measurable	Will not measure to highest accuracy over a very wide frequency range
Resonant circuit	Will measure low loss	Not direct reading; temperature variation difficult; samples must be thick
Cavity resonators	Will measure very low loss	Not direct reading; temperature variation difficult; samples must be machined to shape of cavity (bulky)

listed in *Table 5.2*. The subsequent text will discuss how these circuits operate.

At the lowest frequencies, below 10^{-1} Hz, it becomes excessively time-consuming if not impossible to balance a bridge. Fortunately, the decay of current on applying a steady d.c. voltage '*step response*' can be related to the a.c. loss tangent. Indeed a.c. loss-frequency spectra are the Fourier transform of d.c. current-time. Convenient time intervals (seconds to tens of thousands of seconds) correspond roughly to the inverse in frequency, i.e. 1 Hz to 10^{-4} Hz. The main difficulty lies not in the experimental set up, which simply consists of measurement of current and voltage as in Figure 5.11, but in the

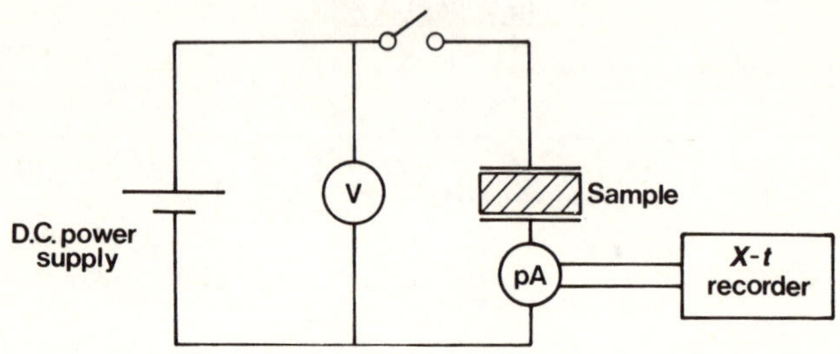

Figure 5.11. Simple circuit for step response technique

theoretical analysis. This calls for complex numerical integrations by computer, but fortunately Hamon derived an approximation for ε'' that is accurate to a few per cent in most practical circumstances. He wrote

$$\varepsilon'' \approx \frac{i}{\omega C_a V} \tag{5.4}$$

where ε'' is the imaginary part of the complex permittivity at frequency f where $\omega = 2\pi f$ and i is the measured charging current after a time t given by $t = 0{\cdot}63/\omega$ seconds. C_a is the capacitance of the electrodes with the sample replaced by air.

Derivation of ε' calls for separation of the current at infinite time (the d.c. current) from the transient current, and a much cruder mathematical approximation, and is still the subject of theoretical investigation. It is therefore common to quote ε'' against frequency as a measure of energy absorption against frequency, since we recall that $\tan\delta = \varepsilon''/\varepsilon'$. Note that $\varepsilon'' = \sigma/\omega\varepsilon_0$ [cf. equation (1.12)]. The Hamon equation is inappropriate when the charging transient is not the mirror image of the decay transient.

For audio frequencies, a simple *Schering bridge* as in Figure 5.12 can be used. Here one adjusts a capacitor and resistor to balance

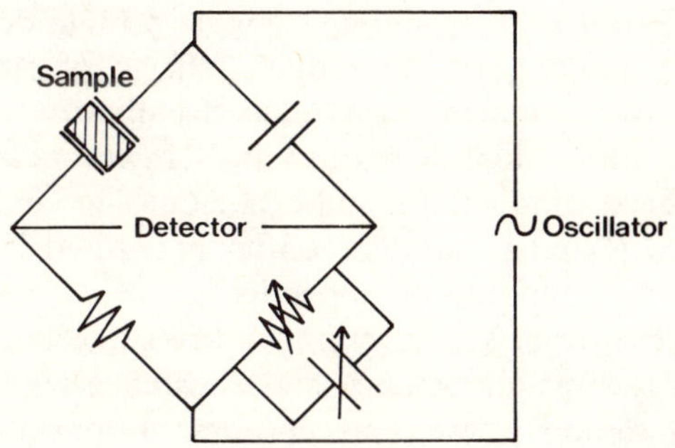

Figure 5.12. Schering bridge

the current in the sample. These can be calibrated to read directly G and C. Range can be varied by replacing the standard fixed capacitor that is shown.

The Schering bridge is particularly versatile. Versions that pass high currents and high voltages (e.g. 100 kV a.c.) are useful in measuring power capacitors and cables, which have high C and hence pass high currents.

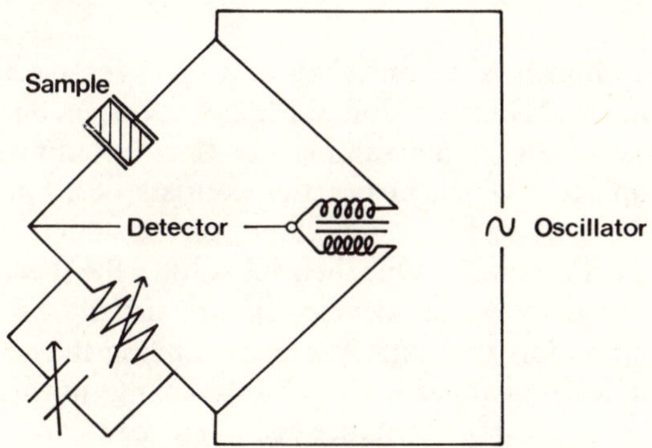

Figure 5.13. Transformer ratio arm bridge

Increases in power factor with time under load may lead to overheating later in life and can be detected at an earlier stage by this method. Schering bridges can incorporate a d.c. bias for measurement of electrolytic capacitors under operating conditions. The Schering bridge is also useful for studying the Garton effect (Figure 3.18).

The *transformer ratio arm bridge* (Figure 5.13) (a development of the Schering bridge) greatly reduces effects of stray reflection between the two balancing arms. It is usually direct reading, and useful up to rather higher frequencies. However, sometimes the equivalent series conductance and capacitance are indicated. These are related to R and C parallel components (used in definition of ε'' and ε') by the equations of Appendix 2.

At radio frequencies, conventional bridge techniques are less useful because stray inductance and resistance tend to be dominating and sample dimensions become important. *Resonance*

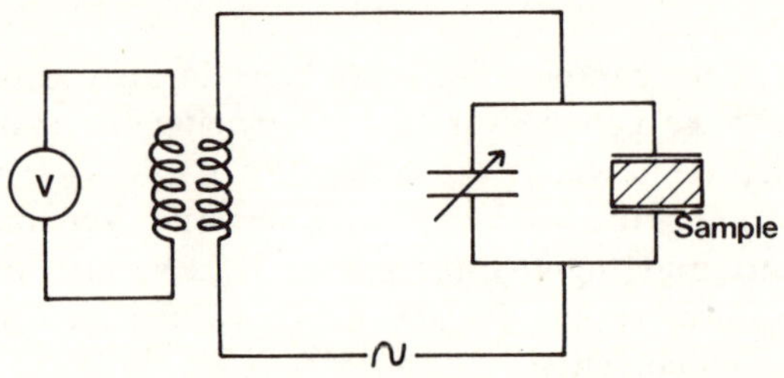

Figure 5.14. Resonance method of measuring dielectrics

circuits are appropriate, consisting of a precision electrode system with minimal leads in the circuit of Figure 5.14. Here one first detects resonance with air in the sample position by adjustment of the variable capacitor which in practice consists of a fine and coarse control. The half width b_0 of the resonance curve is found by adjusting the fine control. One then substitutes the sheet of dielectric to be measured (dielectric devices cannot usually be investigated directly) and replots the resonance curve finding the new half width b_i. The sample capacitance is given by the change in variable capacitance needed to achieve resonance again. Hence ε' can be found from

$$C = \varepsilon'\varepsilon_0 A/d$$

and tan δ is given by

$$\tan \delta = \frac{b_i - b_o}{2C} \qquad (5.5)$$

Hartshorn and Ward designed the most commonly used form of this apparatus.

From 10^8 to 19^9 Hz the Hartshorn and Ward technique is modified so that the sample jig constitutes a closed coaxial transmission line. The supply frequency is varied to achieve the first resonance. Equation (5.5) still applies.

Above 300 MHz, measurement is particularly awkward since the wavelength of the radiation, being less than 1 m, is of the order of the dimensions of the apparatus. Residual inductances cause errors. *Cavity resonators* have to be used where the sample is specially

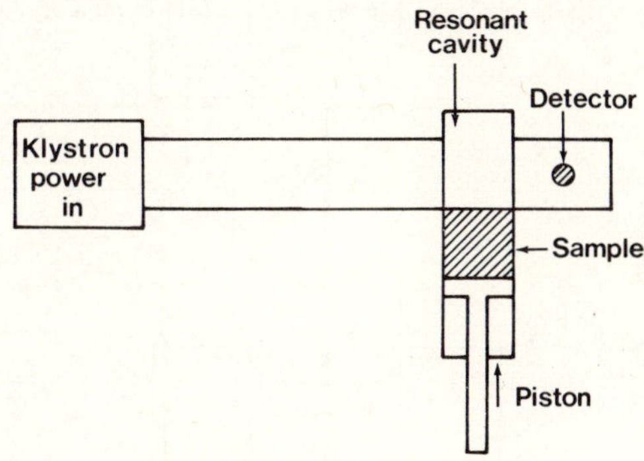

Figure 5.15. Simplified diagram of the cavity resonance technique

shaped to reflect the waveform. The resonance curves are obtained with and without the sample in the cavity by varying cavity dimensions with a piston as shown in Figure 5.15 or by varying the frequency. ε' and ε'' (hence tan δ) are derived from the so-called attenuation factor of the design employed.

5.3.1 THERMALLY STIMULATED CURRENTS

A technique that can be used in the interpretation of time-dependent (a.c.) phenomena is the study of thermally stimulated currents. This is illustrated diagrammatically in Figure 5.16. The various dipoles in the sample are oriented by application of a high field at 20°C or higher, i.e. similar to the preparation of a thermoelectret. The

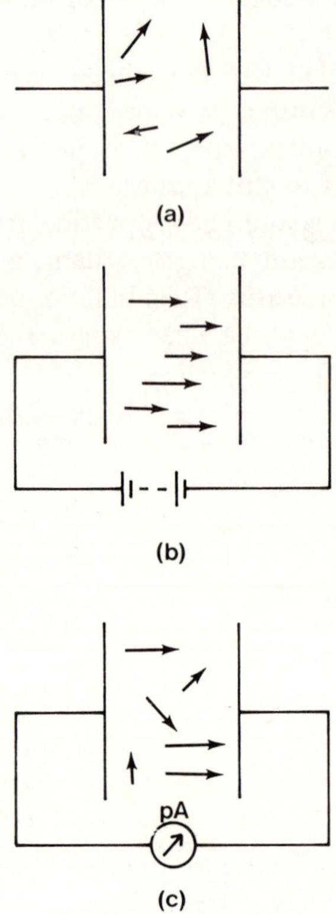

Figure 5.16. Diagrammatic illustration of thermally stimulated currents. (a) Random dipoles; (b) dipoles oriented by field; then sample frozen, then heated at a constant rate; (c) dipole reverting to random configuration at a characteristic temperature and creating current pulse

sample is frozen with the field applied. The voltage is then removed and the sample heated at a constant rate. At given temperatures the various types of dipole will revert to their random configuration and pass currents that may be detected by a sensitive picoammeter.

This gives a current–temperature plot as shown in Figure 5.17.

The advantage of this technique is that, having identified the position of a given peak, one may completely isolate it for analysis by starting the next experiment at a temperature just above. Further, one may identify electronic processes by checking whether ultra-violet light 'bleaches' a peak, and one may examine the energies

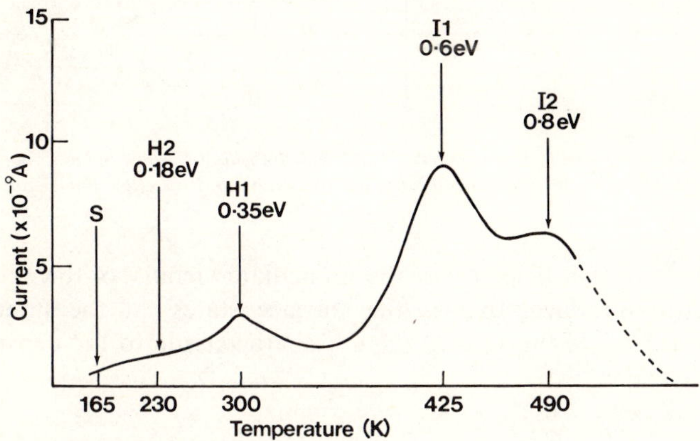

Figure 5.17. The five thermally stimulated current peaks observed in silicon oxide films with their approximate positions and activation energies [After A. Servini and A. K. Jonscher. Thin Solid Films, **3,***341 (1969)]*

involved by using various collecting voltages. Processes other than the orientation of simple dipoles can, however, contribute to thermally stimulated currents. Basically they are allied to electret phenomena.

5.3.2 SURFACE STATES

The usual way of examining the surface states in a gate insulator as used in field effect transistors is to make a silicon–insulator–aluminium sandwich. One then measures the capacitance of this arrangement as a function of d.c. bias voltage applied. Curves like those in Figure 5.18 are usually obtained.

At high positive biases the surface of the silicon touching the insulator becomes depleted of electrons and effectively extends the thickness of insulator, reducing the measured capacitance. The voltage at which this occurs depends on the supply and trapping of

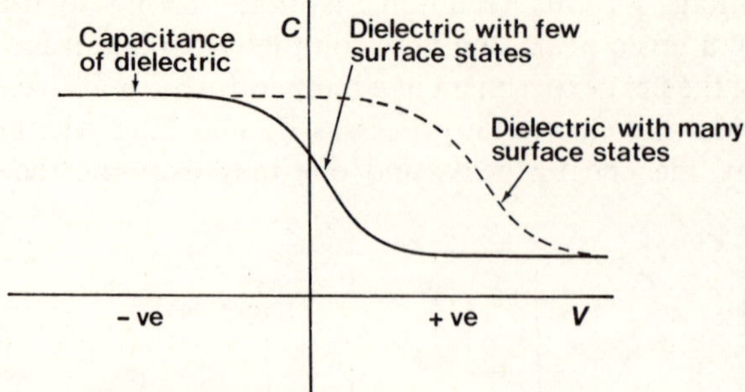

Figure 5.18. Typical C/V characteristic of a metal–insulator–semiconductor sandwich, used in the assessment of gate insulation for application in field effect transistors

electrons in the dielectric in the immediate vicinity of the silicon–dielectric interface, that is the 'surface states' at the interface. Formulae relate the shift of the *CV* characteristic to the density of surface states.

5.4 BREAKDOWN

5.4.1 BREAKDOWN IN SOLIDS AND LIQUIDS

There can be two motives in measuring breakdown. One is to determine the weakest link that will cause failure of an electrical device. The other is to determine the elusive intrinsic strength of the material. In the former case one may use the electrodes and

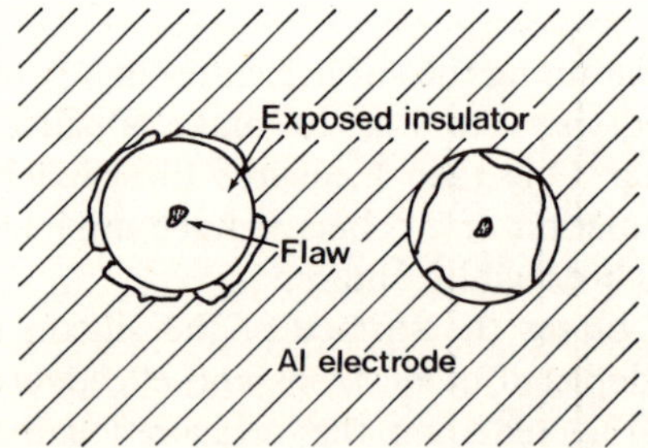

Figure 5.19. Sketch of a typical micrograph showing self-healing action at about ×*600*

environment that are present in the device in question. The breakdown may be caused by a step pulse of voltage, a fast rise as is given by discharging a capacitor, or a slow rise by using perhaps a motor driven variable transformer.

Where sheet or thin-film insulation is used, a higher breakdown strength can be examined by using self-healing electrodes, such as electrolytes, or 0·2 μm metal films that melt under high currents at flaws, isolating them as in Figure 5.19. These are not solely of academic interest as they are used in many capacitors. The usual procedure is slowly to apply a field of the order of 10^8 V m^{-1} to cause self-healing, then to proceed with the breakdown test.

With more massive insulation, potentials of kilovolts must be applied in breakdown testing, and samples immersed in oil and dished as in Figure 5.4 are useful in the avoidance of surface and

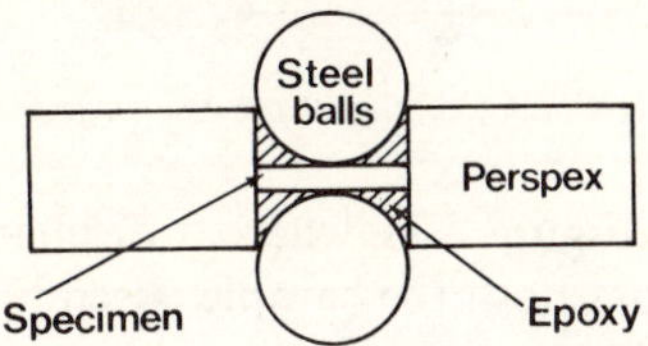

Figure 5.20. McKeown specimen mounting for measurement of the electric strength of solids

gaseous effects. Since discharges around a material can trigger breakdown within it, McKeown cast his specimens in epoxy as in Figure 5.20. Although this technique gives higher breakdown strengths, care must be taken that this is not due to compression caused by the casting process.

5.4.2 BREAKDOWN IN GASES

Breakdown in gases is extremely sensitive to the effects of edges and asperities at electrodes. Spherical electrodes are therefore ideal, although disc shaped electrodes can be used for d.c. if guarded by earthed metal around and a few millimetres out from the edges. Spheres should be separated by a distance no greater than their radii. The situation then approximates to planar geometry.

Of particular practical concern is the detection of gaseous

breakdown in apparently solid, or liquid constructions such as cables, transformers, capacitors and power switches. The weak link in these cases, particularly under a.c., is corona in pre-existing voids or generated bubbles, but glow discharge or spark breakdown may also occur.

The usual way to detect discharges in insulation is from the high frequency noise they generate. This is far more sensitive than using a Schering bridge to detect a final rise in loss tangent with increasing voltage, or listening for sound effects. A suitable circuit is shown in

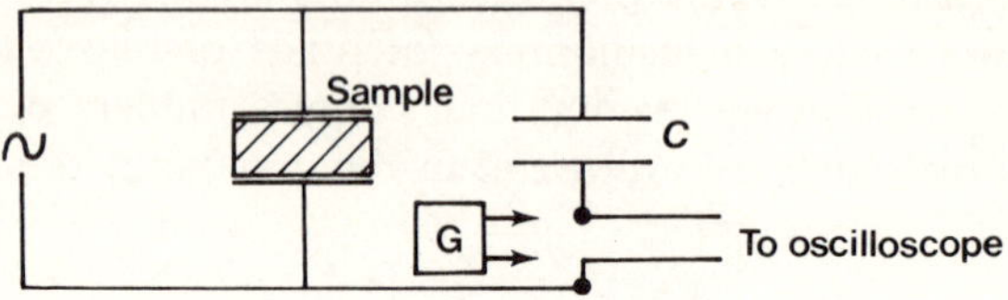

Figure 5.21. Simplified circuit for discharge detection

a simplified form in Figure 5.21 where a discharge-free a.c. power supply is placed across both the sample assembly (say a section of cable) and a discharge-free equivalent capacitor, and the noise monitored. The instrument may be calibrated with a standard pulse generator G. If the output is placed across the Y plates of an oscilloscope and the input across the X plates, a trace as in Figure 5.22 is observed once the 'discharge inception voltage' has been reached.

This allows one to observe the magnitudes of the various types of noise and observe at which part of the cycle they occur. Further

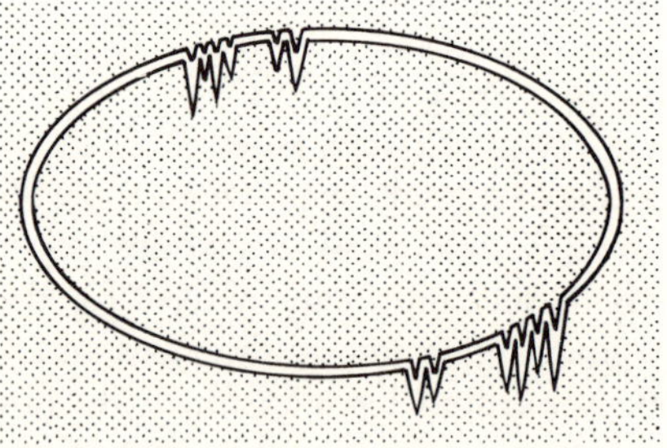

Figure 5.22. Typical oscilloscope trace for discharge detection apparatus

analysis is then possible. In the simplest case, the sample is considered as the capacitor network shown in Figure 5.23. *Table 5.3* gives a guide to the location of breakdown from oscilloscope traces.

Table 5.3 PROBABLE LOCATION OF BREAKDOWN RELATED TO DISCHARGE DETECTOR TRACES

Location	*Position of spikes on trace*	*Time-dependence of magnitude of trace*
Voids in solid dielectric	A few, evenly divided between negative and positive voltage regions	None
Voids in liquid dielectric	As above	Rising
Conductor–solid dielectric interface	Asymmetrically distributed between positive and negative voltage regions	None
Corona in surrounding gas	Close to negative voltage peak only	None
Bad contact	Many small discharges around current peaks	Varies

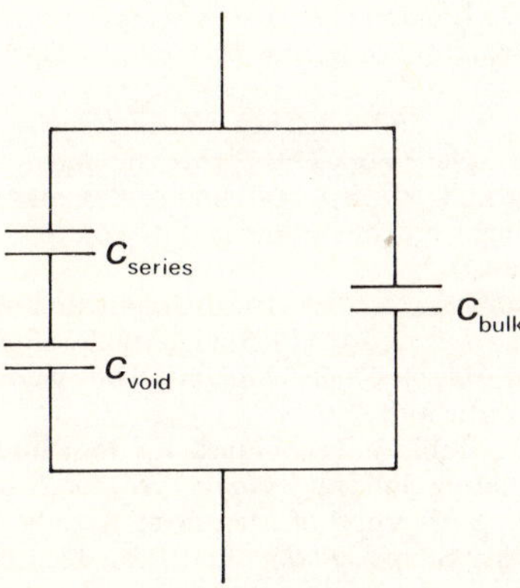

Figure 5.23. Simple capacitor model of a dielectric with voids, as used in the analysis of discharge noise

Great care must be taken with discharge testing, as short-term tests do not necessarily relate to long-term breakdown phenomena experienced in service. The site of the discharge in a large component such as a cable may be detected by shining X-rays on given areas to see if the magnitude of the noise in the detector is reduced.

Impulse testing for corona is also possible. Often initial testing gives an exceptionally high breakdown potential, but after several pulses the value settles at that given by Paschen plot. This may be a fortunate coincidence, since the Paschen plot is strictly only applicable to uniform field breakdown.

FURTHER READING

ALSTON, L. L., *High Voltage Technology*, Oxford University Press (1968) (Measurement of breakdown)

BAKER, W. P., *Electrical Insulation Measurements*, Newnes, London (1965)

DAVIES, D. K., *Proc. Static Electrificiation Conf.*, Institute of Physics, London (1967) (Mobility by decay of surface charge)

DAVIES, M., *Some Electrical and Optical Aspects of Molecular Behaviour*, Pergamon Press, Oxford (1965) (Kerr effect)

DEKKER, A. J., *Solid State Physics*, Macmillan, London (1962) (Hall and thermoelectric effects)

HAMON, B. V., *Proc. I.E.E.*, **99**, iv Monograph 27 (1952) (Low frequency approximation)

HILL, N. E., VAUGHAN, W. E., PRICE, A. H. and DAVIES, M., *Dielectric Properties and Molecular Behaviour*, Van Nostrand, London (1969) (Advanced measurement techniques)

KLEIN, N., TANNHAUSER, D. S. and POLLAK, M., *Conduction in Low Mobility Materials*, Taylor and Francis, London (1971)

KREUGER, F. H., *Discharge Detection in High Voltage Equipment*, Butterworths, London (1964)

MCCRUM, N. G., READ, B. E. and WILLIAMS, G., *Anelastic and Dielectric Effects in Polymeric Solids*, Wiley, New York (1967) (Measurement of a.c. properties)

SERAPHIM, D. P., BRENNEMANN, A. E., D'HEURLE, F. M. and FRIEDMANN, H. L., 'Electrochemical Phenomena in Thin Films of SiO_2 on Si, *I.B.M. Jl Res. Dev.*, **8**, 400 (1964) (Surface states)

SERVINI, A. and JONSCHER, A. K., 'Electrical Conduction in Evaporated Silicon Oxide Films', *Thin Solid Films*, **3**, 341 (1969) (Thermally stimulated currents)

SHEUMAN, P. G., *Diffusion in Solids*, McGraw-Hill, New York (1963) (Radiotracer measurements of diffusion)

SPEAR, W. E., 'Drift Mobility Techniques for the Study of Electrical Transport Properties of Insulating Solids', *J. Non-Cryst. Solids*, **1**, 197 (1969)

VAN DER PAUW, L. J., 'A Method of Measuring Specific Resistivity and Hall Effect of Discs of Arbitrary Shape', *Philips Res. Rep.*, **13**, 1 (1958)

6

Retrospective

Following this broad survey we can now reconsider the meaning of the word Dielectric. Clearly there is no rigid definition. One may broadly categorise a dielectric as a material having the following characteristics, at the temperature and pressure concerned:

(1) Forbidden bandwidth greater than 3 eV.

(2) D.C. conductivity less than $10^{-6}\,\Omega^{-1}\,m^{-1}$ for fields below $10^{7}\,V\,m^{-1}$.

(3) Loss tangent less than 0·5 between 50 Hz and 10^{6} Hz.

This acknowledges that the material may be a semiconductor or metal at other temperatures, frequencies and pressures. However, one must admit that the quantities involved are arbitrarily chosen.

It has been seen that the above criteria are best met by combining elements from the first three periods of the Periodic Table given in Figure 6.1. In practical dielectrics, C, H, O and F most commonly recur plus Mg, Al and Si for liquids and solids, and Cl in gases. However, other useful properties beyond those of simple insulation, in particular ferroelectricity, can sometimes call for the addition of heavier elements. The demons of the piece, that degrade the properties of most dielectrics, are free ions from Groups 1 and 7, particularly Na, K, Cu, Ag, H and, in solids and liquids, Cl.

It is an implication of postulates (1), (2) and (3) that the term semiconductor should be applied to more than the few tens of

Group Period	1A	2A	3A	4A	5A	6A	7A	8			1B	2B	3B	4B	5B	6B	7B	
1																	H	He
2	Li	Be											B	C	N	O	F	Ne
3	Na	Mg											Al	Si	P	S	Cl	Ar
4	K	Ca	Sc	Ti	V	Cr	Mn	Fe	Co	Ni	Cu	Zn	Ga	Ge	As	Se	Br	Kr
5	Rb	Sr	Y	Zr	Nb	Mo	Tc	Ru	Rh	Pd	Ag	Cd	In	Sn	Sb	Te	I	Xe
6	Cs	Ba	La	Hf	Ta	W	Re	Os	Ir	Pt	Au	Hg	Tl	Pb	Bi	Po	At	Rn
7	Fr	Ra	Ac															

Ce, Pr, Nd, Pm, Sm, Eu, Gd, Tb, Dy, Ho, Er, Tm, Yb, Lu

Th, Pa, U, Np, Pu, Am, Cm, Bk, Cf, Es, Fm, Md

Figure 6.1. The Periodic Table of elements—aid in the selection and modification of the dielectrics of the future. Most modern dielectric compounds are composed of elements from the first three periods. Free ions from Groups 1 and 7 are particularly detrimental to dielectric properties

compounds with a forbidden bandwidth below 1·5 eV, as has previously been the case. The upper limit should be extended to 3 eV, since materials up to this limit have most electrical properties in common. Unless the broader definition is accepted someone may invent the term semi-semiconductor! Indeed, the word 'semi-insulator' has been used loosely in this context and needs defining.

Appendix 1

Calculation of Lorentz field

Consider that the dipoles and their associated field are appreciable in a sphere radius R centred on one dipole with terms defined as in Figure A1.1. Then the x direction average field E_x is given by

$$E_x \simeq \frac{e}{4\pi\varepsilon_0 r^2}\hat{r}\,.\,\hat{\imath} \qquad \text{due to the positive charge}$$

$$= \frac{e\cos\theta}{4\pi\varepsilon_0 r^2} = \frac{e}{4\pi\varepsilon_0 V}\int_V \frac{\cos\theta}{r^2}\,\mathrm{d}V \qquad \text{(A1.1)}$$

for the whole sphere of volume V

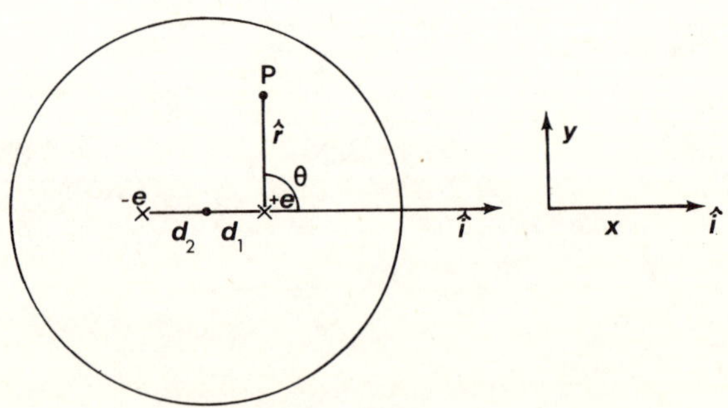

Figure A1.1. Nomenclature for calculation of Lorentz field

For a sphere, $dV = 2\pi r \sin\theta\,(r\,d\theta)\,dr$

$$E_x = \frac{e}{4\pi\varepsilon_0 V}\iint \frac{\cos\theta}{r^2} r^2\, 2\pi \sin\theta\, dr\, d\theta$$

$$= \frac{e}{2\varepsilon_0 V}\int_0^{r_s}\int_0^{\pi} \sin\theta\cos\theta\, d\theta\, dr \qquad \text{(A1.2)}$$

where

$$\begin{aligned} R^2 &= d_1^2 + r_s^2 + 2d_1 r_s \cos\theta \\ &= (r_s + d_1\cos\theta)^2 + d_1^2(1 - \cos^2\theta) \\ &= (r_s + d_1\cos\theta)^2 + d_1^2\sin^2\theta \\ r_s &= (R^2 - d_1^2\sin^2\theta)^{\frac{1}{2}} - d_1\cos\theta \end{aligned} \qquad \text{(A1.3)}$$

$$E_x = \frac{e}{2\varepsilon_0 V}\int_0^{\pi} \sin\theta\cos\theta\,[(R^2 - d_1^2\sin^2\theta)^{\frac{1}{2}} - d_1\cos\theta]\, d\theta \qquad \text{(A1.4)}$$

$$E_x = \left[-\frac{1}{3d_1^2}(R^2 - d_1^2\sin^2\theta)^{\frac{3}{2}} + \frac{d_1}{3}\cos^3\theta\right]_0^{\pi} \frac{e}{2\varepsilon_0 V}$$

$$= -\frac{ed_1}{3\varepsilon_0 V} \qquad \text{(A1.5)}$$

Accordingly, the effect of both positive and negative charges is

$$E_x = -\frac{e(d_1 + d_2)}{3\varepsilon_0 V} = -\frac{m}{3\varepsilon_0 V} \qquad \text{(A1.6)}$$

where m is the electric moment of one dipole; therefore

$$E_x = -\frac{mN}{3\varepsilon_0} \qquad \text{(A1.7)}$$

Now the field in the y and z directions is zero by symmetry. Thus

$$E_i = E + \frac{P}{3\varepsilon_0} \qquad \text{(A1.8)}$$

where E is the applied field, E_i the net field in the material and

$P = mN$, the total polarisation

But $$P = E(\varepsilon'\varepsilon_0 - \varepsilon_0) \tag{A1.9}$$

$$= E\varepsilon_0(\varepsilon' - 1)$$

so $$E_i = E\left[1 + \frac{\varepsilon' - 1}{3}\right] = E\left(\frac{\varepsilon' + 2}{3}\right) \tag{A1.10}$$

which is the usual form of the Lorentz field.

Appendix 2

Relation between series and parallel *R* and *C*

Where a measuring bridge reads the defined parallel properties R and C in terms of a series resistance R_S and capacitance C_S as in

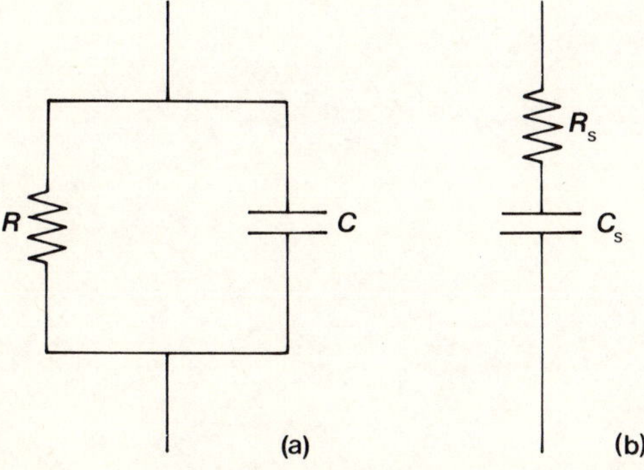

Figure A2.1. (a) Parallel RC circuit defining resistivity and permittivity and (b) equivalent series circuit

Figure A2.1 the quantities are related by

$$R_S = \frac{R}{1 + (\omega CR)^2} \quad \text{(A2.1)}$$

$$C_S = \frac{1 + (\omega CR)^2}{\omega^2 CR^2} \tag{A2.2}$$

or, alternatively

$$C = \frac{C_S}{1 + (\omega C_S R_S)^2} \tag{A2.3}$$

$$R = \frac{1 + (\omega C_S R_S)^2}{R_S(\omega C_S)^2} \tag{A2.4}$$

and

$$\tan \delta = \frac{1}{\omega RC} = \omega R_S C_S \tag{A2.5}$$

R_S must not be confused with the series resistance of electrodes.

Appendix 3

Less familiar terms for quantities used in dielectric theory

Aluminium electrolytic capacitor	Elco, electrolytic capacitor
Capacitor	Condenser
D.C. resistance	Insulation resistance
Dielectric loss angle	Dielectric phase difference
Discharge inception voltage	Corona level (misleadingly)
$\varepsilon' \sin\delta$	Loss index
$\varepsilon' \tan\delta$ (i.e. ε'')	Dielectric loss factor or index
Electric strength	Disruptive gradient, dielectric strength, breakdown strength
ω, ($2\pi \times$ frequency in Hz)	Angular frequency
Permittivity	Dielectric constant, specific inductive capacity
Plastic films	Film
Self-supporting metal films	Foil
Time constant	Insulation resistance wrongly but frequently used; insulance; dielectric quality, insulation quality
Voltage	Tension

L

Appendix 4

Constants and conversions

Avogadro's number $N_A = 6{\cdot}0226 \times 10^{26}\ (\text{kg mol})^{-1}$
Boltzmann's constant $k = 1{\cdot}3805 \times 10^{-23}\ \text{J K}^{-1}$
Charge on an electron $e = 1{\cdot}602 \times 10^{-19}\ \text{C}$
Gas constant $R = 8{\cdot}3143 \times 10^{3}\ \text{J (kg mol)}^{-1}\ \text{K}^{-1}$
Permittivity of free space $\varepsilon_0 = 8{\cdot}85434 \times 10^{-12}\ \text{F m}^{-1}$
Pi, $\pi = 3{\cdot}14159$
Speed of light $c = 2{\cdot}9979 \times 10^{8}\ \text{m s}^{-1}$

Quantity	*Unit*	*Number of SI units*	
Mass	g	10^{-3}	
	lb	0·4536	kilogram
Work (energy)	erg	10^{-7}	
	eV	$1{\cdot}602 \times 10^{-19}$	
	kcal	$4{\cdot}186 \times 10^{3}$	joule
	ft. lbf	1·356	
	Btu	$1{\cdot}055 \times 10^{3}$	
Power	erg s^{-1}	10^{-7}	
	kcal s^{-1}	$4{\cdot}186 \times 10^{3}$	
	ft. lbf s^{-1}	1·356	watt
	Btu h^{-1}	0·2929	
	hp	746	
Charge	emu	10	coulomb

Current	emu	10	ampere
Potential (voltage)	emu	10^{-8}	volt
Electrical displacement D	e.s.u.	$(120\pi)^{-1}$	coulomb m^{-2}
Field	e.s.u.	3×10^4	volt m^{-1}
Polarisation	e.s.u.	$(1/3)\ 10^{-5}$	coulomb m^{-2}
Inductance	e.m.u.	10^{-9}	henry
Resistance	e.m.u.	10^{-9}	ohm
Capacitance	e.s.u.	$(1/9)\ 10^{-11}$	farad
Length	thou or mil	$2{\cdot}54 \times 10^{-5}$	m
	inch	$2{\cdot}54 \times 10^{-2}$	m
	foot	$3{\cdot}048 \times 10^{-1}$	m
	angström	10^{-10}	m

Appendix 5
Glossary of terms used in advanced dielectric theory

The following glossary is intended as an introduction to a few of the terms used in advanced dielectric work.

Bushing	An insulator taking a high voltage conductor through an earthed wall
Cataphoresis	Motion of charged colloidal particles in a liquid under an electric field
Chalcogenides	Compounds, usually two-element glasses, having cations from Group 6B, i.e. O, S, Se, Te
Corona resistance	Time that dielectric will withstand a given level of field-intensified ionisation
Electrostriction	Force due to potential on two electrodes deforming a dielectric in between: common with polymers
F centre	Cation vacancy in a solid filled by one electron, usually imparting a colour to the material
F^1 centre	Cation vacancy filled by two electrons
Factor of assurance	Ratio of test voltage to working voltage for a given device
Formative time lag	Time for a discharge to pass from a low to a high current condition

Frenkel defect	Interstitial vacancy pair created by a lattice ion shaking out of place
Hydrogen bond	The bond between a hydrogen atom (proton) and two electronegative ions such as oxygen. The bond is weak (0·2 eV) and is associated with dipolar and quantum forces
Hyperelectronic polarisation	Exceptionally high permittivity in conductive polymers caused by electrons oscillating along the length of polymer chains
Impulse strength	Breakdown voltage under voltage surges of the order of microseconds
Interstitialcy	A mechanism of ionic diffusion consisting of alternate interstitial and vacancy hops
Microdischarges	Gas discharges of microcoulombs for milliseconds prior to main event
Photodielectric effect	Enhancement of permittivity on exposure to light
Polarisation index	A measure of dielectric absorption usually taken as the ratio of current after 10 min of voltage application to that after 1 min
Pseudoinsulators	Generally applied to materials that are only insulating at high frequencies
Pyroelectric effect	Acquisition of electric charge on heating. Many piezoelectrics are pyroelectric
Rad	Unit of radiation dose usually applied to gamma and X-rays. 1 rad = 100 erg of energy deposited per gramme of material considered
Schottky defect	A thermally created vacancy, where the displaced ion is presumed to have migrated to the outer surface of the material
Thermodielectric effect	The phenomenon by which a potential difference appears between the solid and liquid parts of a two-phase dielectric when the interface moves
Triboelectricity	The phenomenon by which electrical charges are separated by friction between bodies

Index